Kanban from the Inside

Understand the Kanban Method, connect it to what you already know, introduce it with impact

Mike Burrows

BLUE
HOLE
PRESS

Sequim, Washington

Blue Hole Press
72 Buckhorn Road
Sequim, WA 98382
www.blueholepress.com

Printed in the United States of America.

Library of Congress Cataloging-in-Publication Data
applied for; available at www.loc.gov
ISBN: 978-0-9853051-9-2 (paperback)

Cover design by Jane Pruitt
Interior design by Vicki L. Rowland

10 9 8 7 6 5 4 3 2 1 0

❖ CONTENTS ❖

❖ *To Sharon* ❖

❖ FOREWORD ❖

I'm an American who has been fortunate to see a fair amount of the world. In my travels, I've learned that Americans have a unique love of the "Self Help" genre of books: Americans love to be told that they're doing something wrong and that the breathtakingly brilliant book that they're holding in their hands contains the secret code that will make them thinner, faster, stronger, or even more beautiful.

Mike is a Brit, and the world is fortunate that he chose to write this book because the last thing we need is another "Self Help Software Method" in which the author promises to give development organizations the secret code that will make their people write better code that is more aligned to customer, stakeholder, and market needs, all with fewer bugs and the latest technology. Which sadly means that some Americans—and possibly some others suckled in the school of "Self Help"—might avoid this book.

Which would be a terrific shame.

In this comprehensive book, Mike doesn't give us "Self Help" advice. Instead, he gives us something that is vastly more powerful and deeply enduring: Self Insight tools and processes.

As insight is the ability to understand the true nature of a thing, Self Insight is the ability to understand ourselves. From the perspective of this book, the "Self" refers not to ourselves as individuals, but instead to our "Organizational Self"—our company or development team or development organization.

The powerful tools that Mike describes—such as the Kanban board and the Cumulative Flow Diagram—are introduced through the foundation of a set of values that provides a truly satisfying means to realize lasting change.

Perhaps most surprisingly, Mike achieves all of this without demanding, cajoling, coercing, or even asking you, his reader, to do any one thing. Instead, Mike starts with a deep exploration of the values and then shows us how we can act in ways that are congruent with these values. If you "value" these things, you'll find that you want to put those actions into practice. And if you put them into practice, you'll find that you're starting to make the improvements you've been looking for all along.

—Luke Hohmann
Founder and CEO
Conteneo, Inc.

❖ PREFACE ❖

Perhaps you've heard of T-shaped people; well, this is a T-shaped book! It has the depth you would expect in its core subject, and some breadth, too. Its depth, of course, relates to the Kanban Method—what it is, how its practitioners think, advice on how to apply it, and so on. The breadth comes in the form of some important reference points from outside the method. These, I hope, will be helpful both to those looking in from the outside and to those ready to look outward for fresh sources of inspiration.

I set myself the task of describing the "humane, *start with what you do now* approach to change" not as a productivity tool, but as a management method built around a strong framework of values—a way to help organizations work better for their people, their customers, and other stakeholders.

One important feature of this management method is that it can be applied at a wide range of scales, from the level of individuals and small teams all the way up to strategic business initiatives. Don't be surprised to find that we can move from one scale to another in the space of a few paragraphs—the ease with which this is possible is itself a strong statement about how the Kanban Method works.

I'm writing as an experienced manager with a strong technical background. In recent years I've been a global development manager with teams on four continents, an executive director, an IT director, and an interim manager. It has been a while since I was last paid to write code, but it is still a passion of mine, and even now I get drawn into discussions about system architecture and design.

It is only natural, then, that you will find this book a little easier to read if you have some knowledge of software development, but this is by no means a prerequisite. Similarly, I try to make no expectations of familiarity with bodies of knowledge such as *Lean* and *Agile*—we'll come to those as needed. What matters is that you share with me a sense of professional curiosity about managing what we now call *creative knowledge work,* and share my interest in how it might be improved for all concerned.

The book is organized into three main parts:

Part I presents the Kanban Method (or just "Kanban") in a new way, through a system of nine values. I first started writing about the values at the beginning of 2013, and they've been at the heart of my work ever since. Part I closes with some even more recent concepts, Kanban's three *agendas* and the *Kanban Lens.* The agendas are the fruits of a collaboration with my friend and colleague David J. Anderson (the Kanban Method's originator), and the lens is due to David himself.

Part II is all about breadth. Instead of claiming to have all the answers, the Kanban Method includes in one of its core practices the phrase "*using models.*" These two little words encourage practitioners to borrow liberally from other established bodies of knowledge. These include Systems Thinking, Lean, Agile, and Theory of Constraints, but there are many more to choose from. I can't hope to cover all of them in depth, but I do hope that my Kanban-centric perspective on these and other models will lead you to explore them further and to integrate them into your thinking.

Part III describes a repeatable process for implementing the Kanban Method in an organizational context. David refers to this as the "Systems Thinking approach to introducing Kanban" (or STATIK), and it forms the backbone of our foundation-level Kanban training. I've brought it up to date by integrating it with the values and agendas of Part I and explicitly referencing some models from Part II.

You might have noticed already that I use italics to emphasize the deliberate use of a word or phrase as a technical term—*start with what you do now,* for example, is Kanban's first foundational principle; *serious games* refers to something well-defined. You will find many of these terms collected together in the glossary. Terms such as **balance** and **customer focus,** in bold, refer to Kanban's values.

❖ PART I ❖

KANBAN THROUGH ITS VALUES

The approach I take to introducing the Kanban Method is a little unusual. Most typically, it is explained through its *Foundational Principles* and *Core Practices*. I take half a step back from these, and start from a system of nine values. In the order of the first nine chapters of Part I, these are **transparency, balance, collaboration, customer focus, flow, leadership, understanding, agreement,** and **respect**.

This is to take nothing away from Kanban's principles and practices, and every chapter here makes direct reference to one or more of them.

For convenience, I label the four Foundational Principles as FP1–4:

FP1: Start with what you do now.

FP2: Agree to pursue evolutionary change.

FP3: Initially, respect current processes, roles, responsibilities, and job titles.

FP4: Encourage acts of leadership at every level in your organization —from individual contributor to senior management.

Similarly, CP1–6 refer to the six Core Practices:

CP1: Visualize.

CP2: Limit Work-in-Progress (WIP).

CP3: Manage flow.

CP4: Make policies explicit.

CP5: Implement feedback loops.

CP6: Improve collaboratively, evolve experimentally (using models and the scientific method).

Instead of a detailed technical justification for each technique, I hope to convey an insider's view that explains how Kanban practitioners—by which I mean managers, other staff, and external experts who know and apply the method—tend to approach organizational problems. The values help to emphasize the intent of the principles and practices so that you can more easily match them to the needs of your current organizational situation.

The tenth and final chapter of Part I organizes the nine values into three *agendas* and introduces another useful tool, the *Kanban Lens*.

❖ CHAPTER 1 ❖

Transparency

W e're a few floors up in our shiny new office overlooking Budapest's Nyugati Station. It's just gone 9:30 a.m. and our standup meeting is in full swing. Our mood is a little more jovial than usual because we're anticipating a trip across the road for coffee and cake to celebrate our latest "learning." It's my treat—I had managed to leave the build in a badly broken state for several hours the day before, and the IT director isn't supposed to do that.

But we have a more pressing matter to attend to first. Tibor and Máté are expressing concern about a ticket (a yellow sticky note) that has been stuck for several days in the "Released" column on the team's whiteboard. Gy chips in with an offer of help: He is confident that he can convince the Operations team that the new pricing curve represented by that stuck ticket is more reliable than the old one; we should expect it to be fully im- plemented in production in a day or two. Gy's offer is gratefully accepted, and we move on to the next ticket.

Those few moments are enough to demonstrate several examples of Kanban's first value, **transparency**:

Our work is there for everyone to see on a large whiteboard. This board organizes tickets that represent *work items* into columns; these columns correspond to work item *states* ("Released" is one such state) or the stages of our high-level workflow.

We understand the importance of regular *feedback*, and we're holding a *standup meeting*.

We have crystallized some of the *policies* that govern our work; many of these are displayed near the board. One such policy states that developers

3

retain responsibility for their work items until they have obtained customer confirmation that the item is proving its worth. Another policy dictates that learning—most especially the I-won't-be-making-that-mistake-again kind of learning—should be reinforced with the aid of cake.

That scenario played out in 2009–2010, just as David Anderson was seeking feedback on the newly articulated *foundational principles* and *core practices* of the Kanban Method. We—people like me who were already applying the ideas— discussed and refined them through our group mailing list. Within weeks, they went into print in David's "blue book."[1] We had a documented method!

Transparency is central to Kanban. Three of its six core practices relate to it:

CP1: Visualize.

CP4: Make policies explicit.

CP5: Implement feedback loops.

We look at these in turn.

Core Practice 1: Visualize

Kanban's single-word practice, "Visualize," seems a little nonspecific. However, most implementations of Kanban feature a particular kind of visualization, a *kanban board*. These boards are used to implement *kanban systems*, visual work-management systems that have some very useful properties. We explore those properties in later chapters, but for now, we concentrate on the visual aspects.

Had we used an electronic tool, our board might have looked something like Figure 1.1.

Instead of sticky notes on a whiteboard, now we have icons that we can drag around the screen. Whether physical or electronic, these are referred to as the *kanban*, a Japanese word that can be translated as "visual sign," or perhaps, "token." We prefer the more accessible terms *card* (or *ticket*—I use the two interchangeably) and *work item*, leaning to the former when we're thinking about visual design and to the latter when we're focusing on what they represent.

---

1. Anderson, David J. 2010. *Kanban: Successful Evolutionary Change for Your Technology Business.* Sequim, WA: Blue Hole Press.

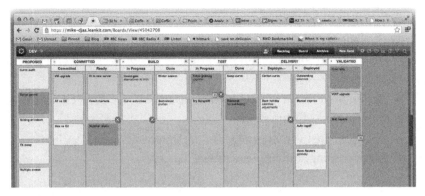

Figure 1.1   An electronic board

In this book, work items represent defined chunks of *knowledge work*, things like product features that need to be developed or service requests that need to be fulfilled. They're not necessarily software-related—we've seen examples in legal, HR, sales, and the CEO's office—what they tend to have in common is that much of the work exists in people's heads or on their computers. Without the board, this work would be invisible.

It might seem a trivial point, but it's important that these tickets are movable. We move tickets from column to column as the underlying work item progresses. This means that we can see at a glance how much work we have in any given state of completion. Try doing that with a to-do list!

If the board is big enough and the ticket design bold enough, we can see all of these things from across the room:

◆ Which work is *blocked* (waiting on something)

◆ Who is working on what

◆ What different types of work we have, and in what proportions

◆ How much work we have at each stage of completion

With all that information constantly available at a glance, we soon get a feel for what a healthy state of affairs looks like. Once we're attuned to it, the board will let us know immediately when things have departed from the norm and some kind of investigation or corrective action is needed.

Visualization and Change

With Kanban, the purpose of visualization and other forms of transparency is twofold: to make the need for action visible and to help people make good choices. These operate at two levels:

◆ Action in the form of work that needs to be done; good choices in the selection of work items

◆ Action in the form of changes to the system; good choices in justifying, scoping, and implementing change

How do you respond when the board indicates that all is not as it should be? Here are some common responses that a manager or coach might make:

1. *Never mind, it will work itself out as it usually does.*
2. *I will intervene by managing things more closely until the situation passes.*
3. *I will intervene by changing the system.*
4. *Do they understand how things got like this? Will they want to make changes to the system? How should I help them?*
5. *I respect their capability to make appropriate changes in this situation.*

All of these responses might have their place, but some appear to reflect higher maturity than others. By provoking action and supporting good choices, Kanban leads people toward higher-maturity responses such as 4 and 5 above. Any organization that consciously embraces these and the styles of **leadership** that encourage them is well on the way to maturity.

Response 5 (and to a lesser extent, response 4) contains another very interesting element, namely *self-organization*.

Self-organization is a beautiful thing. It doesn't just mean that individuals are able to act with autonomy, essential though that is to their well-being. It means also that the system can reconfigure itself to meet its challenges more effectively. Self-organization in this full sense adds adaptability and resilience, and because it can happen without outside intervention, it scales really well. For both system operation and system change, self-organization is both effective and humane.

Changing a system that is managed visually should be a cheap and quick thing to do. Rub out a line or two on the whiteboard, draw another one, and move a few stickies around (or a make few clicks with a mouse).

Given that the impact of these changes might be huge relative to this small implementation effort, working on the system in this way is a high-leverage activity.

There's a virtuous circle at play here:

◆ The kanban system organizes the work.

◆ People organize themselves around the work.

◆ From their fresh perspective, people see that the kanban system could organize the work better than it currently does, and they change it.

Core Practice 4: Make Policies Explicit

Kanban boards are very effective at organizing work, but some system aspects aren't best described in the visual language of tickets, colors, columns, and the like. Sometimes a few words in the form of *policies* can make all the difference. These aren't edicts from on high; they're a way for the participants in a system to maintain a shared understanding of how the system operates.

It's another **transparency** thing—parallel with the strategy of making the invisible visible, we like to make the implicit explicit, but if (and only if) we think it will help us make better and more predictable choices. And again we want leverage—a few words carefully chosen to capture intent, not thick documents that attempt to cover every eventuality. I've seen policies as short as "*Demo!*" stuck above the relevant column on the board—this was all that was needed to reinforce an effective working agreement.

Many policies describe the qualities expected of work items as they enter or leave a column, for example:

◆ *Items in the "Ready" column should require no more than five days' development.* In Budapest we referred to this as the "five-day rule"; later this became a "two-day rule."

◆ *Items can't enter the "Test" column until they've passed peer review and have been demonstrated to the team.*

On or around the board, snippets such as "*< 5 days dev,*" "*Code review,*" and "*Demo!*" serve as perfectly adequate reminders of what the team expects to happen.

Other policies can be more global in nature:

◆ *Production stability takes priority over QA bug fixing; both take priority over new development.*

◆ *When taking on a new piece of work, inform the sponsors if it is likely to have an impact on any existing work.*

These policy examples aren't universal, but they're readily transferrable into comparable contexts. That's not uncommon. Sometimes there's an underlying principle that transfers more easily than the policy—"*celebrate learning*" rather than "*eat cake,*" for example. There is no shame, however, in policies that are completely unique to the situation. Context matters!

Policies and Change

We add policies when we believe that the additional clarity will help us either to make better choices or to make them more efficiently. When we do so, it's often a good time to discuss underlying thinking, such as:

◆ *Larger work items are disproportionately likely to prove troublesome compared with smaller ones.* (This might be more than just a hunch—we might have data that support this hypothesis.)

◆ *Generally speaking, it is better to finish something than to start additional work.* (This could be accepted as a piece of homespun wisdom—"*Stop starting and start finishing!*"—or be based on solid theory.)

When we make policies explicit, we cause that underlying thinking to be tested. If reality doesn't match the thinking, we will keep finding ourselves bumping up against the policy. This creates a discomfort that prompts reevaluation and further learning.

For this reason, it's good to start with simple policies that reflect current practice (what's actually done most of the time, regardless of official policy) and refine from there as necessary. *Start with what you do now* makes as much sense here as it does anywhere. Write them down, then challenge them. Does it always make sense to do a demo here? Might this ten-day piece of development be okay after all?

This strategy—make explicit things that were previously implicit—applies to the definition of the Kanban Method itself. The presence of feedback loops and the practice of implementing them was so "obvious" that no one thought to include it in the method's initial drafting.

Core Practice 5: Implement Feedback Loops

Despite this early omission, feedback loops are essential. Without timely feedback, signs of unhealthiness in the system go unnoticed, get ignored, or get addressed so haphazardly that we're never really sure whether we made a difference. Feedback loops are essential, therefore, to making transparency an effective driver of change.

As with *Visualize*, there is a lot of room for interpretation and imaginative implementation in the wording of this practice. That's deliberate. So, to make it more concrete, let's look at a three common examples of meetings that create regular opportunities for different kinds of feedback. We then finish this section with feedback loops that are based on metrics.

The Standup Meeting

If there's one Agile practice I recommend above all others, it's this one. I've seen teams drop standup meetings when their value goes unrecognized, only to reinstate them just a few days or weeks later as things start to fall apart. Perhaps the very familiarity of the practice makes its value too easy to overlook.

Standup meetings are short meetings—short enough that most attendees could (or should) comfortably stand for the duration—held regularly (often daily). Speed comes with practice: knowing the drill, giving updates at an appropriate level of detail, and having the discipline to take conversations outside.

Standup meetings can take a number of forms:

♦ Informal, agendaless, unstructured (hard to recommend)

♦ Interrogation by a manager (often a project manager asking for updates)

♦ Around the room, person-by-person, perhaps using *Scrum*'s format, in which each participant in turn describes *what I did yesterday, what I plan to do today, and what impediments are in my way*[2]

---

2. The Agile method Scrum is described in Chapter 13.

◆ Reviewing work items on the kanban board, scanning it right-to-left, starting with items that are close to completion, stopping the meeting when it would be unproductive to look any farther upstream

◆ Similarly, scanning the kanban board right-to-left, but discussing only items that are blocked or at risk

The Scrum format and the two board-centric formats all support our background agenda of adding transparency around our choices—what to work on and how, when to intervene tactically to help a piece of work along, and when to step back and look at how the system really works.

There's a social, team-building element, too, of course. Team members benefit hugely, not just from knowing what is going on, but also from the reinforcement the regular updates give to each team member's mental picture of how their colleagues work. As trust is built and the needs of the team become better understood, the more honest and more pertinent will be the communication. In short, it is 15 minutes very well spent.

You might not be surprised to read that I have a strong preference for the board-driven formats. They keep reinforcing the idea that *together* we want to get work across that finish line; they are focused more on the objective and less on the person currently assigned the work. I'm not slavish about this, though, and I find that as teams get comfortable both with the techniques and with each other, we can slide unconsciously but still quite appropriately between different styles, both formal and informal.

Do you hold equivalent meetings? In the spirit of making the implicit explicit, ask yourself: What is the intent behind your meeting format? Is that intent fully realized? Do your team meetings encourage or discourage self-organized problem solving?

Replenishment Meetings

The replenishment meeting is another widely used practice.[3] This is the forum in which a process's *input queue* (or, if you prefer, its *backlog*) is populated with new work. It's also a great opportunity to gauge customer satisfaction, to explore customer needs, and to match those needs with the capability of the team. It's an important feedback loop because it provides an external perspective.

---

3. If you know Scrum, its replenishment meeting is the *sprint planning meeting*.

I'm not going to focus on the mechanics of replenishment meetings for two reasons:

1. They're very context dependent. How you work with a single, internal customer will be very different from how you deal with multiple, external customers, for example. Design the process to fit the situation, and be open to the possibility that an appropriate solution might be meetings held only on an as-needed basis—or no meetings at all.

2. I've seen teams too often fall into the trap of focusing on the needs of the process (or worse, the pet process of its facilitator) at the expense of the customer. Who do all those hoops and hurdles really serve?

Other Meetings

Some meetings exist outside the delivery process, designed to feed or manage the improvement process. A popular Agile practice adopted by some Kanban implementations is the regular team-level *retrospective*, replaced in other implementations by a more immediate "on-the-spot" approach. We strongly encourage larger Kanban implementations to hold departmental *service delivery reviews* (sometimes called *system capability reviews*) weekly or biweekly, and divisional *operations reviews* monthly. In these, multiple teams share performance data, incident reports, and improvement updates with each other and (ideally) representatives of the customer and the wider organization.

You may have these or broadly equivalent meetings already. The goal, of course, is not to have yet more meetings, but to create the conditions in which timely feedback will be generated and acted upon effectively. They are worth designing carefully.

Feedback Loops Based on Metrics

The chart in Figure 1.2 is my *cumulative flow diagram* (often referred to as a *CFD*). I made it by:

1. Every few days, counting how many tickets were in each column of the board

2. Accounting for completed work whose tickets no longer appeared on the board (that's what makes this chart *cumulative*)

3. Visualizing these counts into a stacked area chart in Excel, arranging it such that the completed work sits at the bottom

Nowadays, you might get one of these charts for free, provided by your favorite online kanban tool.

Even if you've never seen one of these before, you can probably make out the "ballooning" of work-in-progress so typical of the beginning of projects, then the "staircase" of big *batch transfers* from stage to stage (releases, mainly), and, toward the end, a more continuous style of delivery.

Yes, there were some significant bumps along the way, but it was good to be able to see that we were winning. I had other ways of knowing that delivery was happening, but this chart gave me clear feedback that *lead times* through the process, *delivery rates* out of the process, and even the style of delivery were being transformed. All this from just one chart!

Other visualizations and metrics help practitioners understand how timings are distributed, giving insight into both the internal workings of the process and the service experienced by the customer.

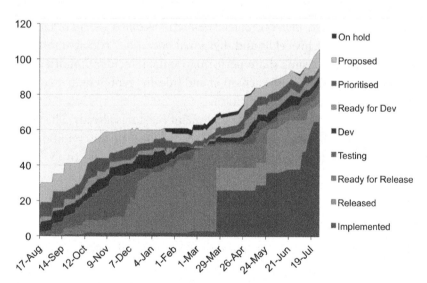

Figure 1.2   My project's cumulative flow diagram

Transparency as a Value

Transparency in Kanban is purposeful—it's there with intent. It's there to help people make good choices about the day-to-day work and about the system. Done well, it promotes self-organization—adding meaning, adaptability, and resilience—and it gives people the feedback that progress is being made. At multiple levels, it prompts underlying logic to be articulated, challenged, and refined.

To the extent allowed by considerations such as **respect**, more transparency means more stuff brought to the surface, more provocation for change. And as organizations mature, we hope that their appetite for transparency increases. More is better, generally speaking.

Not that transparency on its own is enough. Transparency is no substitute for effective **collaboration**; change without **agreement** isn't likely to stick. Sometimes we deliberately combine transparency with other values; for example, with **customer focus** in the replenishment meeting. In short, the key to expanding your organization's limits on transparency may lie in the other eight values.

Chapter 2

Balance

Core Practice 2: Limit Work-in-Progress (WIP)

Balance is a value closely associated with Kanban's second core practice:

CP2: Limit work-in-progress (WIP).

The first part of this chapter shows how WIP limits on a kanban board work as a coordination mechanism for implementing a *pull system*. Later we explore other manifestations of balance in Kanban.

Pull Systems for Knowledge Work

Consider the "TEST" column on the board in Figure 2.1.

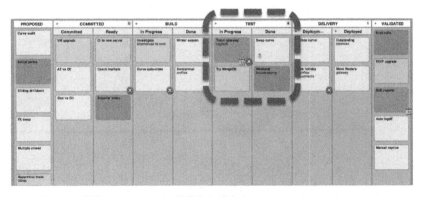

Figure 2.1   The TEST column shows a WIP limit of 4.

At the top right-hand corner of "TEST" there is a "4." This is a *WIP limit*, a limit on the number of items allowed in that area. In this example, the total number of items in the "TEST" column (including its two sub-columns labeled "In Progress" and "Done") should not exceed four. Our tool has highlighted the limit in yellow because this part of the board is currently at full capacity.

Now, look at what happens in Figure 2.2 when tickets move rightward, both within and (importantly) out from this area.

The TEST area of the board is no longer full. The space left behind inside "In Progress" has real significance—it's an availability signal, indicating that we can *pull* a ticket across from a preceding column.

This is how *pull systems* behave: As work moves toward completion (rightward on the board), these availability signals move upstream (leftward), indicating that other work items can move forward into the vacated space as soon as they're ready. Where we have spans of columns with limits (their own or shared), these signals and movements can ripple upstream very quickly—the system appears to be connected, as it properly should. With our kanban board, we have made something quite special—a practical pull system for our invisible commodity, knowledge work.

Without those WIP limits, the pull signals are lost. That leftward signal flow stops as soon as it reaches a column that lacks a limit (here, that's the "PROPOSED" column at the far left-hand side of the board).

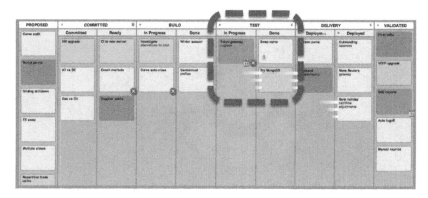

Figure 2.2 Two tickets moving rightward across the board

Those *unbounded* (or *infinite*) queues take on extra significance when we measure the time it takes for work to move through the system. We refer to the time through the WIP-limited columns between unbounded queues as the *kanban system lead time*. This might be very different from the *customer lead time*, the lead time measured from the customer's perspective.[4]

Balance Workload versus Capacity

Consider the board (Figure 2.2) from the perspective of the people most closely responsible for those TEST columns. They could be specialist testers or the same developers or analysts that owned the work earlier in the process; in the interests of self-organization, the board doesn't specify this. One thing the board does make sure of, however, is that the amount of work in this area will not (or at least should not) exceed capacity.

This means that we have taken important steps to avoid the *over-burdening* of people, preventing a situation that is at best unproductive, and at worst, inhumane. As soon as we let finishing take priority over starting, we see quality improve as a result of better focus, and space is created. And with fewer items in any given part of the pipeline, work gets finished significantly faster overall, and feedback comes much sooner.

From time to time, however, it will appear as if the WIP limits have been set incorrectly. Set too high they will seem to have little effect on the flow of work, but if you look closely, it will be apparent that many work items have *stalled*, lacking the people to work on them. Set too low, and too great a proportion of the workload might be *blocked* at any given time, perhaps to the extent that parts of the system suffer from *starvation*, which is a technical way of saying that people find themselves with insufficient work.

These situations should be cause for discussion and investigation, and perhaps for corrective action of one kind or another. The knee-jerk reaction might be to adjust a limit, but don't be too hasty. Make sure that everyone concerned understands the sequence of events that led to the current situation, and work from there.

---

4. Use terms such as *lead time* and *cycle time* carefully. Because their definition can depend on your viewpoint, it is usually advisable to qualify them in some way. It helps, too, if they start at a well-defined point of commitment and end unambiguously with a state of completion, delivery, or validation. Bear in mind also that the technical definitions for several of these terms vary across the literature, and this will continue to generate a great deal of confusion.

WIP limits are best seen not just as mere policy levers, but feedback mechanisms and drivers of system-wide improvement in their own right. When you reduce WIP, you make other problems much more apparent (and yes, they can hurt a little more). Fix those, and WIP can be reduced still further; it may even go down of its own accord.[5] This is another virtuous circle, and a powerful one, famously exploited at Toyota very successfully over a period of decades.[6]

Contrast that virtuous circle with the *vicious cycle* that comes from thinking that the overriding concern should be to keep people busy. With this thinking, the default response to blocked work is not to solve the problem, but to start another piece of work, increasing the amount of WIP in the system. More WIP means more time waiting for other people to finish, compounding the problem. Delays and multitasking both have negative effects on quality, leading to—yes you've guessed it—more rework, more blocked work, and yet more WIP. Too much work coupled with poor quality—is that a combination you recognize?

Other Ways to Limit WIP

I wouldn't want to give the impression that column-level WIP limits are the only way to limit work-in-progress. They're powerful, but sometimes they work best in conjunction with other mechanisms. These tend to fall into one of two main categories:

1. Controlling batch sizes—reducing the size (budget-wise and duration-wise) of projects, the time between releases, and the sizes of sprints, features, and so on

2. Controlling the number of things happening in parallel—reducing the number of business initiatives (to which projects align), the number of concurrent projects per team or department, the number of work items per team or per person, and so on

We have a lot of levers at our disposal here! Push down on the right one, and you'll find others becoming easier—both technically and psychologically—to push down on, too. With WIP constrained in multiple ways, you can bring it down to levels that might once have seemed impossible.

---

5. See *Little's law* in Chapter 17 for a fuller explanation of this effect.
6. The *Toyota Production System* (TPS) is one of the models discussed in Part II; see Chapter 14.

The situation in Budapest with which Chapter 1 opens didn't just arrive overnight. I had joined an organization that was addicted to WIP. As is often the case in young companies, it was very hard to say "no." Unfortunately, this didn't apply just to customer business; not saying "no" had become an internal habit, too. Meeting by meeting, the to-do lists (of which there were many) only ever grew.

A few weeks into my role and already applying Kanban quite conventionally in my part of the organization, a key moment came in a meeting of the management team. Mark Dickinson, our managing director, announced that the to-do lists were history, to be abandoned forthwith. To say that I was pleased would have been a major understatement—I recognized it at the time as a significant breakthrough for all of us.

There were more of these breakthroughs to come. Some months later, we took steps to reduce the number of business initiatives down to a small few that together most effectively supported the business strategy. This in turn had an impact on projects, several of which were put on hold or eliminated completely. Once we had the bandwidth to look at them properly, other projects that had been waiting their turn were found to be very much more urgent than anyone had thought, critically so in one case.

Encouragingly, many of those de-prioritizations didn't wait for me (in IT) to lobby for them—in several cases their sponsors (my peers in the management team) volunteered them. I don't know how often you have seen in-flight projects get cancelled in this way, but in my experience it occurs rarely enough that when you see it happen repeatedly, you know that something very special is happening.

Balance Urgency-Driven versus Date-Driven Work

Not all work is alike, and this is especially true of knowledge work. Many managers try to deny variety or to organize it away, not realizing that it can be far better to embrace it instead.

Let's take a closer look at one small part of our board, as shown in Figure 2.3.

Don't worry too much about the names of the three work items; notice instead the differences in their visual representation. For now, we focus on the top two, "Tokyo gateway upgrade" and "Swap curve."

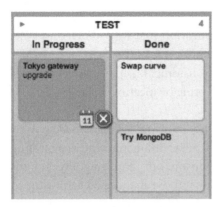

Figure 2.3   The work items in "TEST"

"Tokyo gateway upgrade" has a green ticket annotated with a calendar icon—it is a *date-driven* work item. In this example, the interface to the Tokyo Stock Exchange will change on a particular day, and our systems must be updated to accommodate this. If those changes are delivered late, we won't be able to trade in that particular market, a situation that could be very expensive for the organization. Delivering early, however, yields no benefit—it might even be harmful.

"Swap curve" is different. It represents a completely independent piece of functionality. The sooner it is delivered, the sooner that benefit can be derived from it. Unlike the previous example, it's not date-driven, it's *urgency-driven*.

If we can safely deliver the urgency-driven work ahead of the date-driven work, that's a win, but we'll give priority to the date-driven work as soon as we believe that it is at risk. Either way, we achieve good outcomes.[7]

Treat them the same, though, and bad things happen. If we attach an arbitrary date to the urgency-driven work, we might put the genuinely date-driven work at risk. If we treat them both as urgency-driven, we fail to manage the schedule risk properly. Neither approach encourages good decision making.

Too many teams suffer needlessly by making all of their work date-bound, either by squeezing too much work into time boxes, or worse, attaching dates to work items individually. Estimates become commitments.

---

7. I am simplifying slightly here—date-driven items aren't always so valuable that the desire to deliver on time trumps all other considerations. We look more closely at the economics of decisions like this in Chapter 15.

Productivity targets stretch commitments to breaking point, and that's before the interruptions start to arrive!

Risk-Based Categorization and Classes of Service

"Try MongoDB" (the remaining work item) sounds neither date-driven nor urgency-driven. But does that make it unimportant? It's hard to be sure about this one item, but where will we be a year or two from now if we never risk experimenting, and if we never do the longer-term work behind the scenes to develop our capabilities across platform, product, process, or people?

Looking only at schedule sensitivity (not looking for different kinds of deliverables or different underlying workflows), our examples cover three qualitatively different categories. Kanban implementations commonly recognize four:

♦ *Expedite*: work items so urgent that we will drop other work in order to give them immediate attention.

♦ *Date-driven* (or *Fixed Date*): work items whose delay beyond a specific date will result in a significant penalty being incurred, disproportionate to any benefit in delivering early. Their schedule risks are actively risk-managed.

♦ *Standard* (or *Regular*[8]): urgency-driven work, to be delivered in a customer-agreed order or sequenced according to a system policy. Depending on the context, suitable policies can be as simple as *first in, first out* (FIFO), based on an economic model such as *cost of delay* (Chapter 15), or simply left to personal choice.

♦ *Intangible*: capability improvements, experiments in technology or the market, investments in people—work that's essential over the medium to long term but whose direct and immediate business impact is hard to quantify.

Categorizations based on other risk dimensions are also possible. Your organization's existing planning system might have categorizations of its own already—variations such as *Innovation, Maintenance, Support,* and *Regulatory* are commonplace in banks, for example.

---

8. Marcus Hammarberg and Joakim Sunden recommend the less baggage-laden *Regular* over the much more common *Standard*, and I'd be happy to see it catch on.

When categorizations are based on the need to offer different kinds of performance outcomes, we call them *classes of service* (CoS). Externally, we help the customer choose the class of service that that best reflects the proper balance of risk and reward from their perspective. Internally, scheduling policies guide the choices between work items so that good outcomes are achieved overall. With a healthy mix of work, in particular one that has sufficient buffers of intangible and standard work, a surprising degree of predictability is achievable.

Balance Demand versus Capability

Inside the system, we balance workload against the system's capacity, both for the sake of the people doing the work and for the improved performance and predictability that comes as a result. But it doesn't have to stop there.

Interesting things happen when your system's capability to deliver against each category becomes known. You can help your customers to make better-informed choices. This in turn has an upstream effect, an effect on how work flows from the customer. Managed proactively, this is *demand shaping*, a way to improve outcomes still further by promoting balance across a broader scope of a system and over longer timespans.

Work item categorizations and classes of service help you manage to multiple time horizons simultaneously. There's no point in remarkable delivery rates in the short term if we're busy bankrupting ourselves through insufficient attention to sustainability. Likewise, we can't be forever spending our investors' money on long-term work so nebulous or grandiose that customer value will never be delivered.

In a nutshell, Kanban helps you balance demand versus capability over a range of timespans. This is a powerful management strategy to apply both inside and outside the system.

Stakeholder Balance

Achieving a balance among the interests of different stakeholders—team members, customers, senior management, shareholders, even the wider community—can be especially challenging.

No one has the capacity to weigh up the concerns of all of those stakeholders all of the time, but a practical policy does at least set the right tone: If a so-called improvement works at the expense of any of these groups, think again, try harder. If it feels like a zero-sum game is being played, beware. "*Good for customers, for the organization, and for the people doing the work*" strikes a better tone.

Improvements that don't respect this rule often come unstuck. Customers have only limited tolerance for worsening products and services. Organizations can't usually be expected to support changes unquestioningly, with no regard to their wider impact. People eventually walk away from worsening working environments; productivity, quality, and well being all decline in the meantime.

Seeking Balance

Balance is a strange thing—we really enjoy it when it's there, but achieving it takes anticipation, vigilance, and effort, sometimes even the occasional breakthrough. For me, it is this—as much as the significant technical merits of WIP-limited pull systems—that makes balance such an important value.

To help bring **balance** into your application of Core Practice 2, try prefacing it with "Find ways to":

◆ Find ways to limit work-in-progress, using every available lever.

◆ Find ways to limit work-in-progress at every organizational level, looking for ways to bring those deeper imbalances to the surface as trust is built.

Perhaps you can come up with a phrase that captures the essence of your organization's current need for balance. It's a trick that works with other practices, too.

❖ CHAPTER 3 ❖

Collaboration

We skip ahead now to Kanban's sixth core practice to introduce **collaboration**. Collaboration is a close ally of **transparency** and **balance**; all three are about driving, provoking, or catalyzing change, and their increasing presence is a good indication that things are becoming more sustainable.

Core Practice 6: Improve Collaboratively, Evolve Experimentally

This practice comes with some small print. The full version reads

CP6: Improve collaboratively, evolve experimentally
(using models and the scientific method).

We cover *improve collaboratively* first, then *evolve experimentally* and the *scientific method* together. *Models* are the subject of Part II.

Collaboration Is a Thing

I find it helpful to think of **collaboration** as something quite concrete, bringing to mind some specific examples of famous creative collaborations. Lennon and McCartney, Watson and Crick, Marie and Pierre Curie—these are collaborations that have made huge impact, not just in their chosen fields, but in popular consciousness, too. These aren't just people being nice, cooperating with each other in some general way; these are relationships in which the whole is somehow much greater than the sum of its parts, where creative energy exists between those involved as well as inside each one individually.

We can't expect every workplace collaboration to be quite as spectacularly productive as these examples, but if our knowledge-based organizations can't generate some excess creativity over what each individual can generate on his or her own, why do they exist at all?[9]

Improve Collaboratively

Just as you would expect any effective organization to use collaboration in the delivery of its products and services, *improve collaboratively* simply means directing some of that creative energy toward upgrading the underlying delivery system.

This sheds new light on some of the **transparency** concepts discussed in Chapter 1. *Feedback loops* must involve multiple people. When teams are *self-organized*, improvement and other opportunistic innovation will develop from collaborations that for the most part arise quite naturally, involving appropriate groups of people.

There is no contradiction in saying that we need to work at this, however. Continuous, self-organized improvement doesn't usually just embed itself overnight. Managers sometimes hope that a simple policy announcement will do the trick; almost invariably they are disappointed. You may have seen it yourself: A meeting or two, perhaps a few improvements, then it fizzles out as soon as it becomes clear that this doesn't count as "real work." The next time the appeal goes out, the now cynical workforce isn't remotely interested.

In Chapter 6 we look at the role **leadership** plays in developing these organizational capabilities. Meanwhile, let's look at a specific example of a process change that has collaboration at its core.

Encouraging Collaboration

On a number of occasions, I've seen various forms of *peer review* generate significant frustration. At UBS, we took great pride in the quality of our code, and *code review* (a kind of peer review) was held up as the means

---

9. Why *do* firms exist? Economists still wrestle with that. See http://en.wikipedia.org/wiki/Theory_of_the_firm and http://en.wikipedia.org/wiki/Knowledge-based_theory_of_the_firm

both to "enforce" quality and to spread good practice. In those terms, the practice was pretty successful.

There was a problem, though: Code could get stuck for days (sometimes longer), waiting for the assigned senior developer to find time for the review. Reviewers might batch up multiple reviews "because it's more efficient that way." Significant rework was sometimes required as a result of the review, adding real cost and yet further delay. Multiple delays and rework—this is a recipe for pain.

What was the solution? Collaboration! The formal process that our tooling supported didn't need significant change in this case. It was enough that we agreed that delay and rework would be regarded as evidence that the developer and reviewer had failed to collaborate sufficiently while the code was being written.

This may sound trivial, but the psychological shift here was a big one. Previously, the burden was on developers to get their code past the reviewer. Code review was, in other words, a hurdle. Now, with the informal guiding principle of "no surprises" fulfilling the role of an *explicit policy*, both parties felt a joint responsibility toward the smooth flow of work.

This collaborative approach to the problem of reviews can be taken very much further, and we return to it in Chapter 5.

Focus on Collaboration

One valid lesson to draw from this last example is the simple one that reviews can be problematic—I've encountered it enough times to know that there is a general tendency here. There is a much deeper lesson, though: It pays to look at the *quality* of interactions in a process, not just at how many there are or in what sequence they come.

This is more than just a technique to apply in the context of a specific improvement opportunity—it's a strategy in its own right. Look at the quality of interactions around you. Do you see key processes that are hampered by low-quality interactions, delays, back-and-forth conversations, rework, and frustration? If that's not bad enough, do you see individuals and teams who seem rarely to have the opportunity to engage in high-quality collaboration? Identifying these might be the trigger for something special.

Invest, then, in the skills and practices of collaboration. The Agile community deserves real credit for making this a priority; we focus more on this important contribution in Part II. We return to it again in Part III, showing how some collaborative tools can be used to help establish or refresh the context for Kanban implementations and improvements.

Meanwhile, the next time you are surprised and frustrated by what looks like a failure of an individual or a failure of the process, ask if it could be explained instead as a failure to collaborate effectively. Turn "managers don't like surprises!" into something positive by training yourself and those around you to ask this question automatically.

Evolve Experimentally

If "*improve collaboratively*" is about how change is driven, then "*evolve experimentally*" is about how it is conducted.

Kanban has in common with other frameworks the concept of a problem-solving process applied iteratively in an *improvement cycle*. The Theory of Constraints has, for example, its Process of Ongoing Improvement (POOGI); Six Sigma has the DMAIC loop. There are many more—describing their improvement cycles seems to be a favorite way for frameworks to differentiate themselves.

The Kanban community is happy to stick with the granddaddy of them all, the canonical improvement cycle known variously as the *Deming Cycle*, the *Shewhart Cycle* (which is what W. Edwards Deming himself called it, naming it after Walter A. Shewhart, whose ideas he went on to develop), the *PDCA Cycle*, or just plain *PDCA*. The acronym is short for "*Plan, Do, Check, Act*"; sometimes you will see *PDSA*, for "*Plan, Do, Study, Act.*"

A note of caution: The words *Plan, Do, Check*, and *Act* don't mean what you think they mean—they're even a little misleading—until you use them with the word "*experiment*":

♦ **Plan** an experiment (based on a hypothesis).

♦ **Do** (conduct) the experiment.

♦ **Check** (or **Study**) the results (or outcome) of the experiment.

♦ **Act** on the results, changing either the hypothesis or the system accordingly, sharing appropriately.

It takes some discipline to conduct change in this scientific way, but it certainly adds new meaning to "results-oriented"!

Them and Us

You might say that a team that has solved a problem is a stronger team. That's usually something to celebrate, but there is a point beyond which the positives of collaboration are offset by some negatives: cliques, competitive behavior, information hiding, unfair assumptions about the motives and intentions of those outside, and "bubbles," whose walls become barriers.

This seems like a hard problem, one deeply rooted in human nature and surely outside the domain of the Kanban Method. Or is it?

Kanban practitioners don't generally rush to confront these negative team behaviors directly; neither do we seek early battles with team composition or their members' sense of team identity. Instead, we take a sideways approach, choosing to look at work in ways that don't emphasize organizational structure and individual roles. You can take that as a useful design guideline as you implement **transparency**, **balance**, and **collaboration**, but this surprisingly effective strategy becomes more explicit as we explore the remaining six values.

REFLECTION:
TRANSPARENCY, BALANCE, AND COLLABORATION

We've seen the first three of the nine values, and it's time to take stock. Step back for a moment and reflect:

◆ What practices tend to drive change from within your organization? What intentions, assumptions, and values lie behind those practices?

◆ How readily does your organization embrace **transparency**? Is it seen as a threat, treated with indifference, or valued as a positive force?

◆ Does your organization achieve **balance**? Does it even try? How do imbalances affect teams and individuals? How many can you identify?

◆ Does your organization value **collaboration** enough to nurture it, to make it a specific focus for improvement?

Which of the following interventions would be most beneficial in your current environment?

◆ **Transparency**:
  ◆ Making work visible
  ◆ Making policies explicit
  ◆ Implementing feedback loops (or making better use of feedback loops that exist already)

◆ **Balance**:
  ◆ Implementing a pull system
  ◆ Implementing risk-based classes of service
  ◆ Finding a healthier mix of work (by type or time horizon)

◆ **Collaboration**:
  ◆ Encouraging collaborative problem solving
  ◆ Taking steps to improve the quality of interactions
  ◆ Structuring improvements as experiments

Where would you start? Part III (Implementation) will give you some ideas if you're unsure.

❖ CHAPTER 4 ❖

Customer Focus

Core Practice 3: Manage Flow

Is this a mistake? How do we get from *Manage Flow* to **customer focus**? Indulge me for a moment—let me cheat a little, expanding the wording of this core practice to express more fully what this practice really means:

> **CP3** (*expanded*): Manage flow, seeking smoothness, timeliness, and good economic outcomes, anticipating customer needs.

This chapter focuses on customer needs and how to anticipate them better. Smoothness and timeliness are covered in the next chapter, on **flow**. Keep in mind "good economic outcomes" as you read both chapters; economic decision-making is covered in Chapter 15.

Why Customer Focus?

Task focus, role focus, team focus, project focus, product focus, company focus, technology focus . . . the list goes on. So many ways to lose sight of what we're in business for!

In my classes I offer this advice:

Know what you're delivering, to whom, and why.

You might think that this could go unsaid, but it really seems to hit home. Unsolicited, students tell me that they'd never really thought about it before. They mention it in feedback sheets as a key takeaway. It's not that those other focuses are bad, but that customer focus helps to put them all into proper perspective.

In this chapter, we explore some practical ways in which customer focus can improve the flow of work, sometimes profoundly. It's not so surprising when you think about it: If you look at what you do from a someone else's perspective, you are likely to learn something new about how it works.

Satisfaction Assured

Recall this *policy* from the scenario that opened Chapter 1:

♦ *Developers retain responsibility for work items until they have obtained customer confirmation that the item is proving its worth.*

This policy was a relatively late addition. We had evolved a development process that seemed effective enough. We'd gather requirements, build new features, test them, and release them. After a while, we got a little more sophisticated: We added a column on our board that let us track features that were released but still required further implementation steps before they could be considered complete.

Too often, though, when we checked, we found that we'd delivered features that would never be used. Features that had been asked for! How does that happen?

Our new policy was added to address what we assumed at the time to be bad customer behavior. Why ask for stuff you don't need? How about letting us know when you change your mind? However, it soon became apparent that this new policy was changing behavior on both sides. Closing a *feedback loop* was the catalyst for a level of *customer collaboration* (a value straight out of the *Agile Manifesto*[10]) not previously seen.

Knowing that the process was going to end in what could turn out to be a difficult conversation, developers and our internal customers alike made sure to nail those final implementation steps (clarifying timetables, keeping people suitably informed and trained, getting static data cleaned up, and so on). When necessary, these steps would be tested beforehand, often collaboratively. That, in turn, influenced the way development and specification were done. All the way back at the start of the process, it changed even the way work got prioritized, now that it was apparent that success depended on shared commitment.

---

10. See Chapter 13 for more on the *Agile Manifesto*.

I'm not exaggerating when I say that the impact of this policy change went way beyond my expectations. Some humility is in order too: We didn't have bad customers, just relationships that weren't effective enough.

Right Across the Board

Our catalyst was a policy attached (figuratively speaking) to the right-hand column of our kanban board, customer focus somehow infecting the whole process. To understand how a transformation like ours might be repeatable, it is helpful to review the board with some specific questions. Working right-to-left, column-by-column:

◆ Whose needs are explored in this stage of the process, and how? Whose aren't, and what risks does that pose? [11]

◆ What do we learn in this stage that we don't (or can't) know earlier? In what ways do the activities of this stage help us home in on what will be needed?

◆ What is still to be learned? Are outstanding uncertainties best dealt with by pressing on or by going back?

Working that logic all the way back to the start of the process, no longer are we building to meet given requirements, but building to meet needs that are still to be discovered and explored. Not forever looking backward, justifying ourselves, proving that we are building "correctly to spec," but looking forward, working toward meeting needs that are still unfolding. Neither do we put undue trust in a supposedly watertight process to do our work for us; instead we seek ways to capture learning more effectively.

Creative knowledge work isn't just about what we already know. It is (and I use a piece of technical jargon quite deliberately) a *process of knowledge discovery*. Use your kanban board to keep reminding you: "What don't we know?"

---

11. Don't forget internal stakeholders here—audit, security, finance, support, and so on. If they hold a veto, they should in some sense be treated as customers. See Middleton, Peter and James Sutton. 2005. *Lean Software Strategies: Proven Techniques for Managers and Developers.* New York: Productivity Press.

Upstream Kanban

Let me give you the flavor of one of the more ambitious board designs that I have used to manage my personal workload. The key feature of this design involves the three columns under "IDEAS." Notice how their descending WIP limits seem to suggest a funnel.

IDEAS			COMMITTED		
Priority 3 15	Priority 2 8	Priority 1 5	Ready 4	In Progress 3	Complete

Figure 4.1   An example Personal Kanban board design

This kind of design is helpful when you're as concerned with organizing the ideas and tasks you could be doing in the future as you are with managing your current workload. There are two key tricks to operating it effectively:

1. You train yourself to mentally blank out the leftmost columns until it's time to replenish or reprioritize columns farther to the right. Soon this becomes an unconscious habit, and there's absolutely no doubt that you're operating a pull system. When you are ready to replenish the "Ready" queue, you'll be able to do that entirely from the relatively small number of items (just five) currently in "Priority 1," which is much easier than looking at the entire pool of ideas.

2. You balance your eagerness to add new ideas to the board with a willingness to remove items that are never likely to make it to the "Ready" queue. Removed items can be archived for later review or completely destroyed (I like to move individual items quickly to a holding area "below the line" and deal with them periodically en masse).

WIP limits are, of course, the reminders to practice these two personal disciplines. Once the board starts to fill up, you will find yourself depriortizing work much more often. This is healthy!

My inspiration for this design is the *priority sieve*, a *Personal Kanban*[12] technique. It might seem odd to digress into Personal Kanban now (we return to it briefly in Part II), but I use this design as a model for *Upstream Kanban*, a name coined for the practice of operating a kanban system upstream of the delivery process. Upstream Kanban is about organizing needs and developing ideas so that there are always good choices on offer when delivery capacity becomes available.

This design reinforces two concepts that the portfolio managers of some of our largest corporations seem to forget:

◆ We generate more ideas than we can possibly use —in fact, we'd have reason to worry if it were otherwise. Over time, we will accumulate more ideas than we can usefully manage, let alone implement.

◆ Ideas cannot proceed on their own merits alone—they are in competition with others. Moreover, new ideas can enter the competition at any time.

Something special happens when items move into the "Ready" column: This system has a very unambiguous *commitment point*. To the left of it, we'd be happy to see items moved backward or discarded—that's the system fulfilling an important part of its purpose. To its right, we'd think that something had gone wrong if items failed to progress with good speed, and we'd regret the waste of effort if items were abandoned before completion.

Just as it does inside the delivery process, effectiveness upstream depends on the values we've explored so far:

◆ **Transparency**: The system must make visible the difficult choices that need to be made. The decision-making rationale should itself be explicit. Decisions are the focus of feedback loops (prioritization meetings, for example).

◆ **Balance**: The amount of WIP in the system is controlled, both to maintain a reliable supply of high-quality ideas and to force timely decision making. If needed, additional control can be gained by allocating WIP by customer, budget line, risk category, strategic initiative, and so on.

◆ **Collaboration**: The work of qualifying items for further development is shared among the originators of those items and the peo-

---

12. Benson, Jim and Tonianne DeMaria Barry. 2011. *Personal Kanban: Mapping Work | Navigating Life*. Seattle: Modus Cooperandi.

ple who will service them. Instead of sucking risk into the system prematurely, all parties (and there may be several involved) keep their options open until commitment is timely.

Let's see what **customer focus** adds:

◆ Whose needs do we think are met by these ideas?

◆ Are we meeting needs fast enough?

◆ What is the data telling us? What are people telling us?

◆ What might lie behind those needs?

◆ What needs might be going unmet?

◆ How can we test that?

In short: Can we develop a better sense for what will be needed?

Anticipating Needs

If there's a single idea that I'd like you to take away from this chapter, it's making the mental shift away from doing what is asked, taking orders, fulfilling requests, meeting requirements, and so on, and reorienting the process toward discovering and meeting needs. It's a shift from an internal perspective (what we think we know) to an external one (what's still out there to be discovered). It's also a shift from the past (what we've been told) to the future (when the customer's need will be met).

That emphasis on the future is captured very nicely in the closing words of the Toyota Customer Promise. I found this displayed on a plaque behind the customer service desk at my local Toyota dealership:

. . . anticipating the mobility needs of people and society ahead of time

Think of a service on which you personally rely. Wouldn't you be delighted if the provider anticipated your needs ahead of time? What innovations might they need to introduce in order for that to happen? Can you translate that kind of thinking into your workplace?

❖ CHAPTER 5 ❖

Flow

As our team first took shape, only a minority of members had previous experience with any defined process, Agile or otherwise. Half of them were new to the technologies we were using—even the programming language was unfamiliar—and had never encountered techniques like continuous integration and version control before. Even so, they went on to produce applications of significant size and complexity, developing as they did so a robust process capable of delivering new features of known quality to their customers in a matter of hours or days.

How was this transformation possible? Our approach wasn't to train the team in some new process or to ask people to adopt new roles. It was to apply Kanban, the *start with what you do now* method. Quite simply, we organized our work visually, kept on noticing our most serious impediments to **flow,** and strove to eliminate them.

Core Practice 3: Manage Flow (Again)

Here, again, is my expanded version of Kanban's third core practice:

CP3 (*expanded*): Manage flow, seeking smoothness, timeliness, and good economic outcomes, anticipating customer needs.

We look at managing flow for *timeliness* toward the end of this chapter. The bulk of this chapter is devoted to *smoothness.*

Smoothness

Kanban isn't unique in its appreciation for smoothness. In some circles, it's much stronger than that: It's an obsession! Manager training at Toyota includes many hours watching production lines, looking for the minutest deviation from the smoothness ideal. The strong implication here is that there is no production process so smooth that further improvement isn't achievable and worthwhile.

The idea of staring at production lines doesn't translate very well into the domain of creative knowledge work, even when we're practicing Kanban. All the same, we soon come to feel real discomfort when we see work interrupted, blocked, left unfinished, or sent back for rework—even when the business circumstances prevailing at the time seem to justify these conditions completely. If that pain prompts the question, "What is it about our process that lets this happen so easily?" we've taken an important step forward.

In his book *The Culture Game,*[13] Daniel Mezick reminds us that we have the capacity to *pay explicit attention* to only a limited number of things at once. Whether or not we realize it, when we design and equip our work-management systems, we are making choices about the things we will consciously attend to. In bringing **transparency** to the flow of our work, we have made one such choice. To borrow a line from Mezick, "*Kanban is paying explicit attention to flow.*"

What Does Flow Look Like?

Here are some indications that work is beginning to flow:

◆ You see a good number of work items advancing between standup meetings. Exactly how many and how often is a function of work item size, team size, and the interval between meetings, but for most hands-on work it's reassuring to see some movement every day. Learn how to break larger work items down into smaller (but still valuable) items if progress doesn't seem very visible.

◆ Relative to the number of people available, you see a good number of items free of blocking issues, giving you confidence that progress can and will be made.

◆ You sense (and perhaps measure) that work is finishing faster and more predictably. For work of similar size and type, lead times are typically shorter and they fall within a narrower range.

13. Mezick, Daniel. 2012. *The Culture Game.* Guilford, CT: FreeStanding Press.

If you know you're not seeing these, you have something to aim for. If you think you are, ask yourself whether flow can be improved further by any or all of these broad measures—fully expecting that they can be.

Although it's good to be able to see and even quantify flow, let's not underestimate the importance of *feel*. It feels good to have work items progressing, to have a workload that is not dominated by issues, to be able to make predictions with some confidence. And it feels better for everyone—customers, the organization, and the people doing the work.

Right Across the Board (Again)

The previous chapter encourages you to review your workflow with respect to **customer focus**. Our examination started at the point of customer satisfaction and worked its way carefully upstream all the way back to the original customer need, examining each step through the lens of knowledge discovery. We can try the same approach for **flow**, but with a different lens.

Why work backward like this? It forces us to focus on what we really need in order for work to flow, rather than on enumerating the many ways we could use everything that happens to be at our disposal at each point. Metaphorically (and perhaps even physically), we're standing at the right-hand edge of our kanban boards and thinking about what tends to stop work from being pulled across smoothly to completion, instead of standing at the left side and coming up with ways to push work faster into a growing pile in the middle of the system. Kanban's protocol for standup meetings is a thinking tool in disguise!

What will our new lens look like? When we're paying close attention to flow, we're asking questions like these of each column:

◆ How do work items leave this stage in the process? By what criteria do we know that they're ready? How are those criteria expressed? How is this readiness signaled so that it can be acted upon downstream?

◆ Typically, how much time do work items spend in this stage? How much (if any) of that time is spent in active work, as opposed to just waiting?

- ◆ What are the most significant sources of unpredictability—are they in the work or in the waiting? Do items typically wait for internal availability, or for external dependencies to be resolved?
- ◆ How much of this stage's capacity is absorbed in rework? Or in *failure demand*,[14] which arrives only because work previously completed failed to meet customer needs adequately?
- ◆ How do work items arrive into this stage? How do we know that they're ready to be worked on?

Caution: Questions like these already assume the following:

1. That the process has sensible objectives—to address the right kind of needs
2. That the workflow is scoped sensibly—it starts with the right kind of questions and finishes with the right kind of results
3. That the workflow is organized sensibly—sequenced to generate high-value learning as quickly as possible

I've seen enough counterexamples to know that these can be dangerous assumptions. To take just two:

- ◆ I once helped to straighten out a process whose objective was the approval or rejection of design change proposals. It became apparent that this so-called change management process was disconnected from any actual implementation process; its objectives and scope were either poorly chosen or misaligned. Needless to say, any customer satisfaction delivered out of that process was somewhat short-lived.
- ◆ Project-based software development processes often operate on the assumption that the time to start thinking about *acceptance testing* is toward the end of the project. This delays the opportunity for some high value learning to very late in the process.

There Is Always a Bigger Context

We've already seen how **customer focus** invites teams to consider a context that is broader than their own work. **Flow** does the same, and in multiple dimensions.

---

14. See Seddon, John. 2003. *Freedom from Command and Control: A Better Way to Make the Work Work*. Buckingham, UK: Vanguard Consulting Ltd.

In attending to flow, we often find ourselves forced to consider how things on which we depend—material, information, ideas, even people—arrive at our part of the system. Things tend not to arrive in good time by accident. Neither can we expect our work always to leave our control in good time if we haven't thought beyond our own four walls to make sure that the exit route is clear.

You may have realized by now that Kanban isn't really about individual or team productivity. Yes, we are often working to improve the system within our span of control, but we don't limit ourselves only to problems that we can solve on our own. Often we're reaching out, working with other people on bigger, more end-to-end problems. **Collaboration** here doesn't just help us get things done, it helps to ensure that our collective efforts are focused where it matters most.

Some Low-Level Examples

What follows is a partial list of improvements we made in that 2009–10 period. Some of these might seem very specific to software development, but most contain aspects that do transfer in some way to other domains. They are arranged in "right-to-left" order, working backward through the delivery process:

- ◆ As described in the previous chapter, we added a customer validation step at the end of the process. Our original motivation was to speed the transition from "merely released" to "actually implemented." With hindsight, the real impact of this change is that it catalyzed a specific kind of customer collaboration that was focused on ensuring good outcomes. This brought about significant reductions in end-to-end times, rework, failure demand, and abandoned work.

- ◆ Automated releases. What initially took several hours of intensive manual effort with a high risk of error could now be done in a few minutes with high reliability, and we had the option of a fast rollback in the event that something looked wrong.

- ◆ Automated testing. We started with low-level *unit testing*, but increasingly we automated much of our acceptance testing too. We learned to review test code before reviewing the application code, and drove a number of substantial improvements to the codebase from that direction. This, in turn, didn't just improve the quality of the tests that we'd just written, it made future changes easier to test too.

◆ The "Demo!" policy—requiring that each new development be demoed to the whole team (not just the reviewer) before work moves from "Development" to "Test." This prevented some work from progressing prematurely, and it created good opportunities for knowledge sharing and idea generation.

◆ We adhered to the principle of "no surprises" around code review, but with a modern twist. We deemed code review to be unnecessary when code had been developed through *pair programming* (intensively collaborative programming, two developers working together with one screen).

◆ The "5-day rule"—an agreed policy that stated that work items should represent no more than 5 days' estimated development time. Later, this became the "2-day rule." We refined this policy with an exception (invoked only sparingly) to cover work that could not easily be decomposed.

◆ Daily, rather than weekly, rotations onto support. With our old weekly support rotation, development work regularly suffered from long interruptions, seriously impacting our predictability. As the new daily system took hold, we evolved policies around the morning handover of support tickets to ensure that sufficient help was on hand when the support burden was high (intelligently balancing the flow of support work against the flow of development work).

◆ We looked closely at where support requests originated, negotiating local policies for those parts of the business that generated the most work. The support burden went down as common problems were eliminated and business teams shared learning internally.

Flow Across the Larger Organization

As organizations grow larger, the more their flows involve multiple services. Typically, some services are dedicated to a single flow while others are shared across flows, perhaps outsourced. Some are understood to be part of a "main" flow; some make their contributions only on request.

Figure 5.1 illustrates some of the problems of scale.

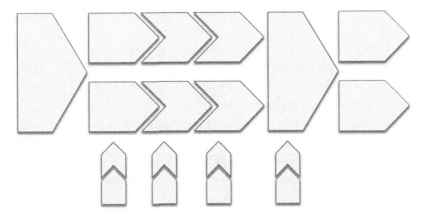

Figure 5.1 Shared and dependent services supporting two main flows

There are many opportunities for delays and frustration at this kind of scale.

- The group responsible for receiving requests for work is different from the one that eventually delivers it. Fire-and-forget communication (email especially) leads to work falling between the cracks.

- Mismatches exist between one person's or one team's idea of "Done" and another's idea of "Ready." These quality problems can be failures of policy or failures of collaboration; often they're both.

- Activities take place on the wrong side of an organizational boundary, away from the most appropriate skills or sources of information.

- The process has unnecessary steps, steps that could easily be absorbed into others, or steps that ought to be split out and brought forward so that important feedback can be generated sooner.

- Inadequate attention is paid to the shared services on which the main flow depends. This problem works both ways: One service's dependency problem is another's upstream problem, the latter a problem of both visibility (requests come too late from people who don't necessarily understand what they are asking for) and capacity management (requests are made without knowledge of the overall workload).

Work is managed in batches that are far too big; for example, multi-month or even multi-year projects that deliver no customer benefit in the interim. Compounding this problem catastrophically, these big batches are started with little thought given to the impact on existing work (other big batches).

It's not unusual to see these problems in combination. The result isn't just extremely poor performance organizationally, but real pain at the human level. Who wants to be constantly overburdened and yet never seem to achieve anything meaningful?

Some will tell you that when things are this bad, you throw it all away and start again. It's ironic: The same people who would champion incremental and evolutionary approaches to product development seem only too eager to recommend disruptive and revolutionary change in people-based systems—in which the outcomes are so much less certain.

The good news is that it is rare to find a process so complex that problems such as these just listed can't be identified and addressed sequentially. The following tactics are often surprisingly straightforward, and they all have obvious relationships to Kanban practices:

♦ Tracking work end-to-end and at the right work item granularity, often at multiple levels of detail

♦ Controlling batch sizes with WIP limits and with policies on work item size—institutionalizing a diminishing appetite for delivery commitments fixed far into the future

♦ Managing WIP across functional boundaries

♦ Agreeing on and collaboratively maintaining policies across functional boundaries

♦ Visually managing dependencies on shared services; conversely, managing demand upstream from these services

♦ Changing the granularity of the activities or states through which work items are managed (again, at multiple levels of detail, if necessary)

♦ Incrementally reorganizing the process, allowing functional reorganization to follow in its wake as required

♦ Measuring and reporting performance in ways that are relevant to customers (many of these will be measures of flow)[15]

---

15. In some parts of the community, *Manage flow* was once worded *Measure and manage flow*. This was never to my taste; measurement is important, but this wording gave it too much prominence. In time, the addition of *Implement feedback loops* (CP5) rendered this old wording obsolete.

Organizations often get deeper into trouble because people think that the answer is to keep trying harder. They think that better project management will fix a problem of capacity management (scapegoating project managers, meanwhile), that stronger functional management will improve end-to-end performance (when it can easily make it worse), or that people should just try to do better (when the system is fundamentally unreliable).

Kanban's unusual solution to this problem isn't to address head-on the roles of project management and functional management; certainly it does not set out to replace them immediately with other things. Instead, it gives managers (and others, too, of course) the tools to see work and how it flows in new ways, together with controls on WIP that impact dramatically on problems of delay and unpredictability. Allowed the right scope, the improvements and the necessary new thinking grow hand-in-hand. If roles then need to change, fine!

Scope should not be taken for granted. Typically (and unsurprisingly), we get brought in as consultants to advise or train a functional area that is under strain, often the most overburdened part of the system. We have to remind our new sponsors that there are limits on what can be achieved with internally-focused improvement efforts, even when they're based on Kanban.

To be frank, an internally-focused Kanban initiative is a neutered one. In particular, it lacks the method's outward-looking values of **customer focus** and **flow** (along with **leadership**, explored in the next chapter). These provide direction to the more inward-looking values and practices of the early chapters. Whether it is Kanban-based or otherwise, if a change initiative close to you is running out of steam, could attention at the right organizational level to one or more of these values help get things moving again?

Managing Flow for Timeliness

It would be a mistake to think that managing flow is only a matter of removing impediments as they arise. Typically in knowledge work, we see work items vary widely in both content and value, and we see the overall workload varying greatly over time. This means that there will always be a place for managing work proactively:

♦ Unusual risks and dependencies must be identified early and managed effectively.

♦ Looking toward the medium term, anticipated workload must be met with adequate capacity.

♦ Longer term, entirely new capabilities might be needed.

Being committed to self-organization doesn't mean that you must always shy away from managing the most important work items more carefully. Not all work items are alike, and some are considerably more deserving of management attention than others. Dates—meaningful dates, at least—should be met. And when it's justified by the business opportunity, it's perfectly valid to sacrifice a little predictability and allow high-value items to jump the queue.

That said, the mark of a reasonably well-evolved and well-calibrated pull system is that most day-to-day scheduling decisions are easy to make. You can let these happen naturally, while at the same time developing the knack of knowing what items to push, and when. As soon as you are able to formalize how you do that, you can help the system evolve further. As the system matures, these interventions will be necessary less often.

❖ CHAPTER 6 ❖

Leadership

Leadership—one of Kanban's most distinctive values—is best described in terms of the other eight values. Through the first five, namely, **transparency, balance, collaboration, customer focus**, and **flow**, we have introduced all six of the Kanban Method's Core Practices. In their conventional order, these are

CP1: Visualize.

CP2: Limit work-in-progress (WIP).

CP3: Manage flow.

CP4: Make policies explicit.

CP5: Implement feedback loops.

CP6: Improve collaboratively, evolve experimentally (using models and the scientific method).

Each of these practices is full of leadership opportunities.

Still to come are the values of **understanding**, **agreement**, and **respect**. These *leadership disciplines* stand in for the original set of Kanban's Foundational Principles:

FP1: Start with what you do now.

FP2: Agree to pursue evolutionary change.

FP3: Initially, respect current processes, roles, responsibilities, and job titles.

These principles are foundational in two senses: They keep the Kanban Method firmly grounded, both technically and philosophically, as an evolutionary approach, and they describe some important commitments and behaviors of leadership that are essential for its initial introduction and ongoing operation.

Leadership is the bridge, therefore, between these two sets of values, binding the practices and principles.

Foundational Principle 4: Leadership at Every Level

The fourth Foundational Principle reads

FP4: Encourage acts of leadership at every level in your organization —from individual contributor to senior management.

Notice how inclusive this language is. It respects your organization's current structure, no individual is excluded, and it imposes no false dichotomy between leadership and management. But don't underestimate its power—it is a strong statement. As a Foundational Principle, it suggests that any failure to attend to this *at every level* kind of leadership may leave your Kanban initiative seriously undermined. But this presents a challenge: What if this kind of leadership doesn't come naturally to your organization?

Fortunately, Kanban doesn't leave you to solve this problem on your own. When change is stimulated, it creates leadership opportunities both large and small. The more widespread, repeatable, and visible this process is, the more positive its impact on your organization's culture will be.

Opportunity Everywhere

Let's test that. Where can we find these leadership opportunities?

♦ **Transparency**: In knowledge work, things don't make themselves visible or explicit by themselves; leaders choose to make them so. This is as true in the small details—the wording of a policy, for example—as it is in the bigger things, such as institutional feedback loops.

♦ **Balance**: Where are we overloaded, and why? Are our pain points obvious, or does the volume of work hide them? Is the mix of work right? There is leadership opportunity in asking these questions as well as in the decisions that may follow.

- **Collaboration**: Making an introduction, reaching out, sharing a problem, noticing how people interact—all of these can be acts of leadership.
- **Customer focus**: It takes leadership to acknowledge that the process may be ineffective at discovering and meeting real customer needs.
- **Flow**: Are you seeing it? What is stuck today? Where do blockages repeatedly occur? Why is that? These are everyday questions of leadership.
- **Leadership**: Encouraging leadership in others can demand real leadership on the part of the encourager. Kanban's kind of leadership not only spreads, it reinforces itself.

Leadership Begets Leadership

My Budapest story illustrates that last point.

Early in our evolutionary journey, we instituted the Agile practice of *retrospective meetings*, in which we gathered periodically to reflect on our recent experiences. In the beginning, we did not find the meetings very effective. One apparent impediment was my presence as manager. We were finding out the hard way that manager-facilitated retrospectives are something of an anti-pattern; I had to step back somehow.

A cascade of "small acts of leadership" followed:

1. I bow out, suggesting that the next meeting be held while I'm out of the office (this was easily scheduled—I was working alternate weeks in Hungary and at home in the UK).
2. Krisztian volunteers to facilitate the next one, starting right away with some pre-meeting preparation. The meeting takes place in my absence and without my involvement.
3. I arrive back in Budapest to find our board annotated for the first time with some written policies. Remarkably, these include some policies that are stronger than I could have expected to introduce myself without significant push-back from the team.

Seeing the enhanced board made me feel that it was no longer just mine—it was the team's, *ours*. And it didn't stop with the board. For example, I might try to move work to the "Testing" column only to be asked, "Mike, has your code been reviewed?" and I would have to admit that it had not. The team had taken ownership of the process to such an extent that they had the confidence to challenge even the manager. Moreover, that newfound process ownership represented a new level of concern for quality.

Why Value Leadership?

Stories such as these show that there is no contradiction in valuing leadership and yet simultaneously encouraging the kind of *self-organization* described in Chapter 1. They need not be in opposition to each other.

It hasn't escaped my notice, however, that some people and some communities struggle with leadership—perhaps they've experienced too much bad management, or they find leadership difficult to square with notions of equality. Let's deal with those.

To the first point, bad management is a real problem that needs to be called out. However, to condemn all managers at a stroke would be not only lazy, but it is incompatible with at least two of Kanban's values—in particular, **understanding** and **respect**. Practitioners of the humane, *start with what you do now* method must be more disciplined than that.

To the second point, leadership can be open to anyone if the environment doesn't actively squash it. People can choose whether or not to exercise it (and that's okay); the real burden is on those who occupy positions of authority to allow leadership to flourish. Note, however, that weaker expressions such as *delegation* and *empowerment* aren't adequate replacements—too often they are so tightly circumscribed that they do not represent meaningful opportunity.

To be clear, we don't value leadership because we value hierarchical organization structures and management styles. They hardly need our blessing! If anything, Kanban's *at every level* kind of leadership will be an awkward fit for some of these cultures. The reason we value leadership is that the relationship between leadership and change is just too strong for a method concerned with change to ignore. Change can be spontaneous, even accidental, but mostly it is the fruit of leadership. And leadership that doesn't produce change is what, exactly?

What Does Leadership Look Like in Kanban?

Earlier in this chapter we listed just a few of the ways that the first six of Kanban's values point to opportunities for leadership and change. It's worth reviewing them one more time in order to get at least a partial picture of what Kanban suggests organizations need from leaders. This is particularly for those in formal leadership roles, but not exclusively so. You need no special badge in order to be a leader in Kanban!

Together, **transparency**, **balance** and **collaboration** suggest a need for leaders who are committed to sustaining an environment in which opportunities for change are easily recognized and systematically followed through. These leaders seek better outcomes for all stakeholders, helping to create the conditions in which people will work together on the ever-present challenge of improving the system.

With the values of **customer focus**, **flow** and **leadership**, leaders make sure that change is carried out in its proper context, with direction and purpose. Leaders realize that some of the most important measures of success are external rather than internal. They appreciate the central role of flow in serving the customer more effectively, in improving the safety and well-being of the people inside the system, and in benefiting the bottom line for the wider organization. The best leaders are in the business of growing leadership in others, and they do not limit their investments to only the areas they control.

That's just the start. What conditions are needed if that kind of leadership is to thrive? The next three chapters cover the remaining three values, namely the *leadership disciplines* of **understanding**, **agreement**, and **respect**. These get to the heart of the Kanban Method's approach to change.

REFLECTION:
CUSTOMER FOCUS, FLOW, AND LEADERSHIP

Imagine going around your organization, asking the question, "*What are you delivering, to whom, and why?*" What sort of answers would you expect?

Would the phrase "*satisfaction assured*" ring true everywhere?

If the lack of smooth flow causes pain, where would your organization be hurting the most? Who would notice, and how?

How well do these words describe changes you have recently encountered (or have been responsible for)?

- edict
- campaign
- cascade
- big bang

- distracting
- irrelevant
- imposed
- random

Where these apply, can you think of some words or phrases that you wish would apply instead?

Imagine an unannounced visit by a senior executive at your usual place of work. You are asked these three questions:[16]

1. What is the process?

2. How can we see that it is working?

3. How is it improving?

What is your reaction? What could be going on here?

---

16. These are in the style of a Toyota *leadership routine* or *kata*—see Chapter 14.

❖ CHAPTER 7 ❖

Understanding

The **understanding** value goes with Kanban's first Foundational Principle.

FP1: Start with what you do now.

Given this chapter's title, you would be forgiven for wondering if this principle should have been stated as

Start with understanding.

As nice as this sounds, this would have been a mistake. It might have been taken to mean "*Don't change anything until you've completed a big current-state analysis project*" or "*Start by aligning your thinking with mine.*" We intend neither of those.

With Kanban, just a little of the right kind of understanding is enough to catalyze the desired reaction. After that, we expect to see understanding continue to develop as "kanbanization" takes hold and kanban systems and their associated practices are implemented and encouraged to spread and evolve.

This focus on *what you do now* is quite deliberate—it's about keeping change anchored in present reality, both now and as it continues to develop. **Understanding** represents both the initial commitment and the ongoing discipline to maintain that anchor's hold. It's quickly actionable (Part III explains how), and yet it contains also the promise of something deeper—it's an open invitation to explore Systems Thinking and its related models (see Chapter 11).

Change without Understanding—
Three Anti-Patterns

Three influential figures have reinforced the crucial relationship between understanding and change. They identified three management anti-patterns— common patterns of management behavior that belie a deficit of understanding, each of which can lead to serious consequences.

- ◆ Russell Ackoff bemoaned *complacency*,[17] the sin of management inaction.
- ◆ Jim Collins, in his book *Good to Great*,[18] warns against *bravado*, management recklessly taking for granted the organization's ability to change.
- ◆ W. Edwards Deming—whom we first met in Chapter 3 and we meet again in Chapter 11—spoke often of *tampering*, a management tendency to address unpredictability in ways that actually make it worse.

Beware of complacency, bravado, and tampering: change that is too slow, too fast, or too random!

Complacency

Fairly or otherwise, it is easy to blame management complacency when organizations are overtaken by events. Spectacular failures are, however, comparatively rare. What about slow decline, inattention to underperformance, or failure to recognize or deal with harmful behaviors (at whatever level), and so on?

Complacency is insidious. It is difficult to deal with because it is hard to acknowledge in oneself; and the list of areas in which one could be accused of being complacent is very long indeed.

Bravado

Stories of bravado aren't hard to find. This one recently made the front page of the Companies & Markets section of the *Financial Times*:

---

17. Ackoff, Russell L. 1991. *Ackoff's Fables: Irreverent Reflections on Business and Bureaucracy.* Hoboken, NJ: John Wiley & Sons.
18. Collins, Jim. 2001. *Good to Great.* New York: HarperBusiness.

G4S Blames scandals on overly aggressive acquisition strategy

G4S's new chief has blamed a short-term and over-aggressive acquisition strategy for a string of scandals at the security group.

A day after the Serious Fraud Office opened a criminal investigation into UK Government contracts held by G4S, Ashley Almanza, chief executive, said: "I don't think as a management team we have always focused on core values and, at times, the short-term focus has undermined those."[19]

I'm not close to this story and I don't know how it will play out, but it seems possible that the company's very survival could be at stake. Sadly, this is by no means a unique or uncommon situation. In my own career I've seen first-hand what happens when a large employer is shaken to its core by the action or inaction of its own leadership.

In 2008, the global financial services giant UBS—at that time my employer of almost a decade—wrote down nearly $38 billion in losses on US subprime debt. That's a difficult number to grasp, but to put it into some perspective, it's $5 for every man, woman and child alive on this planet and substantially more than the estimated $30 billion annual cost of a program proposed by the United Nations that same year to end world hunger.[20] Even to a Swiss bank accustomed receiving depositors' cash in quantities its global competitors can only dream of, $38 billion is a lot of money to lose.

Well aware that it was severely weakened, UBS had already received an injection of capital from the Government of Singapore Investment Corporation (GIC), and it went on to receive further assistance from the Swiss National Bank. Fast-forward to the present: Its workforce has shrunk by thousands, and its share price still languishes at less than a quarter of its pre-crisis peak.

UBS wasn't just an innocent casualty of global economics. What was particularly troubling about this painful situation—and believe me, it was very painful for all who lived through it—was the degree to which the bank had engineered its own near-destruction.

---

19. *Financial Times,* Nov. 6, 2013.

20. "UN says solving food crisis could cost $30 billion," *New York Times,* June 4, 2008. http://www.nytimes.com/2008/06/04/news/04iht-04food.13446176.html

Acting on a benchmarking study by external consultants, it had embarked on a period of rapid growth to strengthen its market position in areas of relative competitive weakness. It made a big push into mortgage-backed securities, of which the subprime market forms a part. Uncharacteristically for a Swiss bank, its cash-fuelled growth in business and staff far outpaced its risk-management capability. Its exposures grew so out of control that its subsequent fall was almost inevitable.[21]

I'm no fan of benchmarking, but here I'm faulting neither that nor the bank's challenging objectives. Challenge can be motivating, unifying—even noble. The problem was change-related: Change was allowed to outrun the organization's knowledge and capability. When individuals and organizations are put at risk, *bravado* is a safety issue.

Most of us don't get close to failures on quite the scale as UBS's, but examples of bravado aren't hard to find: delivery dates plucked out of the air; careers, family life, and health put on the line for the sake of projects that are of dubious value; failed change initiative followed by failed change initiative.

Tampering

Tampering has been defined as "*over-reacting to variation,*"[22] which I much prefer to the narrower and more technical definition, "*adjustments made in response to common-cause variation that result in additional variation.*"[23] It seems to me that tampering should be defined more by the human tendency (over-eagerness to intervene) than by its results.

Perhaps these sound familiar:

♦ "*This must never be allowed to happen again,*" in response to something slightly awkward for the manager and unlikely to be repeated
♦ "*Henceforth . . . ,*" the prelude to an edict

21. This series of events was well documented in April 2008 in UBS's own report to investors, Shareholder Report on UBS's Write-Downs http://www.static-ubs.com/global/en/about_ubs/investor _relations/agm/2008/agm2008/invagenda/_jcr_content/par/linklist_9512/link.277481787.file/bGluay 9wYXRoPS9jb250ZW50L2RhbS91YnMvMvZ2xvYmFsL2Fib3V0X3Vicy9pbnZlc3Rvcl9yZWxhdGlvbn MvMTQwMzMzXzA4MDQxOFNoYXJlaG9

22. *Variation* is discussed in Chapter 11. For the purposes of this section you can read this as "unpredictability" or "unexpected events."

23. Source: http://curiouscat.com/management/deming/tampering

◆ *"This reorganization will deliver the following benefits . . . ,"* heralding another swing of the pendulum and another round of upheaval
◆ *"Something must be done,"* mainly to ease someone's conscience

It's easy to be cynical about these things, but in knowledge work (and in corporate life, generally), unnecessary unpredictability is arguably the least of tampering's costs. I see two greater costs:

When individual decisions are repeatedly taken out of people's hands, they lose autonomy. At the extreme, people feel micromanaged, even unsafe. In other words, tampering has a direct human cost.

When distant managers issue edicts as streams of uncontextualized changes to be implemented, the result is what might be called "policy accretion." Processes become increasingly constrained, which reduces innovation, adds delays, and significantly increases overall economic costs. These aren't the hallmarks of increasing process maturity; they're the signs that sclerosis is setting in. The economic cost is long lasting.

Introducing the J Curve

If your organization is adopting Kanban, it is already attacking complacency (not that any method or tool can be a complete cure). It is beginning to make the need for change visible, and it should soon have some data to show that it is moving in a good direction. If it is allowed to operate at the scope that it needs to, Kanban is naturally anti-complacency.

To address the other two anti-patterns, bravado and tampering, some additional discipline is required. A useful way to visualize this is the *J curve*. We return to it in the next chapter and look at one noteworthy analysis of the J-curve effect in Chapter 17; for now let's just consider its general shape (Figure 7.1 on the next page).

The vertical axis typically refers to quantities such as *performance, capability, fitness,* or *comfort* (all four of these work fine in this context). The J curve traces the impact of a change over time. Its shape suggests three questions:

1. **Will we end up better off than we were at the start?** How sure can we be that we will enjoy better performance, capability, fitness, or comfort after the change?

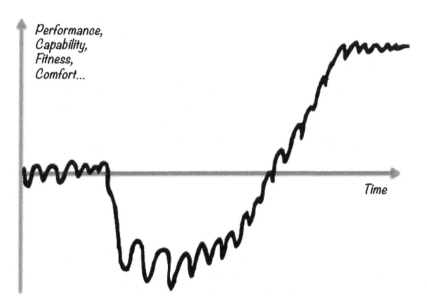

Figure 7.1   The J curve

2. **Will we endure the transition?** Is the intervening loss of performance, capability, fitness, or comfort sustainable?

3. **Do we have the patience?** How much time can pass before the
   urge to abandon the journey will be irresistible, another initiative
   fails, and the change agent gets fired?[24]

The more hastily conceived or grandiose the change, the more likely it
is that at least one of these questions will have a negative answer.
My friend Andy Carmichael puts it like this:

Evolutionary change takes us from survivable point to survivable point.

To which I reply:

And the journey must be survivable too!

You've survived this long already; some appreciation of *what you do
now* must be worth something. So, too, must be some knowledge of the
surrounding terrain (some idea of the possible effects of moving away
from what you do now) so that a safe direction of travel can be established.

---

24. . . . or promoted! I owe the "change agent gets fired" line to David.

A Pattern for Purposeful Change

If that was too abstract, let's make the **understanding** principle more concrete:

FP1 (expanded): Start with what you do now, understanding

◆ The purpose of the system

◆ How it serves the customer

◆ How it works for those inside the system

◆ How it leaves customers dissatisfied and workers frustrated

◆ How it can be changed safely

This contains practice as well as principle—it's actually a highly distilled summary of STATIK,[25] the implementation approach described in Part III. It works both for Kanban's initial introduction and for subsequent changes.

Understanding the purpose of the system takes us from *what we do now* to why we serve our customers in the way we do. For my former team in Budapest, it took us from "*We build and support energy risk-management systems*" to "*We build and support the systems that enable our company to help our customers manage their energy risks.*" For a team that would struggle to get past "*We are software developers*" (and we definitely weren't one of those), understanding purpose could be a big step.

We take the dissatisfactions and frustrations as indications either that there are obstacles preventing the system's purpose from being fulfilled, or that we don't yet understand that purpose well enough. Acknowledging them, their possible root causes, and their effects is a key first step toward making impactful change.

In Budapest, our customer and team frustrations were mirror images of each other, suggesting a common root cause. Quite understandably, our (internal) customers wanted their needs met sooner. On the development side, we were frustrated by changes in priority and by delays that seemed to originate on the customer side. We needed to work together more closely somehow.

---

25. Strictly speaking, it summarizes STATIK-0: "Understand the purpose of the system" is step 0 of STATIK's "advanced mode."

As outlined in previous chapters, we implemented a rapid succession of changes over a period of months that progressively addressed these frustrations. And they were all safe enough—all were reversible, and none generated significant anxieties (certainly not on the scale of the frustrations they were addressing). This was *evolutionary change*, the deliberate strategy that goes with our next value, **agreement**.

❖ Chapter 8 ❖

Agreement

Kanban's second Foundational Principle is very direct:
FP2: Agree to pursue evolutionary change.

In just a few more words: Agree that change is necessary; agree to pursue it with an evolutionary strategy.

Pursue Evolutionary Change

The word *pursue* is very well chosen. So much stronger than *adopt*, it evokes the energy and tenacity to counter any *complacency*, and it reminds us that it's an ongoing process. Pursuit combines challenge and commitment.

A strategy of deliberate evolutionary change cannot mean, "Let nature decide." Rather,

◆ It starts (and keeps on starting) with where we are now—what we do now, the competitive landscape, and so on.

◆ It is open-ended—it is not about following a plan toward a designed end point whose suitability can only be guessed at.

◆ It accepts that any change (environmental as well as internally generated) may cause impacts and generate responses that cannot be foreseen; change is unavoidably *complex* and dynamic.

◆ It is the pursuit of fitness, not "change for change's sake." Fitness may be defined in absolute terms, but ultimately it is fitness for purpose relative to the competition that will determine survival.

◆ Indirectly, it is about *adaptability*. The pursuit of fitness involves repeated change; whether or not it explicitly works at this (and I believe it should), any organization that pursues evolutionary change is developing an adaptive capability, the ability to respond to change.

Start with what you do now, again and again, indefinitely, relentlessly. Your competition won't let up, so neither can you; the ideal end point (if ever there were such a thing) will change as the landscape changes.

We can take the J curve from the previous chapter and chain a series of them together (see Figure 8.1).

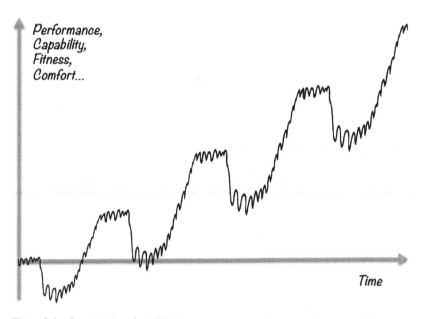

Figure 8.1   A succession of small J curves

Figure 8.1 makes two major simplifications compared with what is typically experienced:

It is unlikely that every change will succeed. With that in mind, only in the most extreme situations would it be appropriate to attempt changes that are likely to inflict lasting damage in the event of failure. In normal circumstances, changes should be *safe to fail*. Perhaps the picture should look more like Figure 8.2.

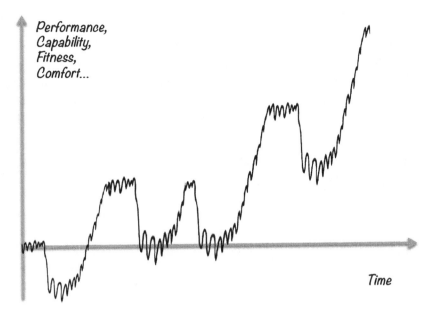

Figure 8.2   Not every experiment will succeed.

In practice, these *safe-to-fail experiments* aren't usually implemented with an "experiments-in-progress limit" of just one. Running diverse experiments concurrently maximizes the chances of finding something unexpectedly valuable.

As an aside: These strong parallels between evolutionary strategies also apply to product development—lots of diverse experiments, ideas generated by a diverse group of people. Like self-organization, diversity adds resilience, and diverse groups solve problems faster. The parallels should not be surprising—the service is, of course, part of the product.

Agree to Pursue

In Chapter 5, I asked you not to take for granted the scope of your Kanban implementation. Now I'd like to remind you that the **agreement** principle (FP2, in other words) starts with the word *agree*. Don't take for granted the people involved, nor those likely to be impacted by it.

By way of analogy, I'm reminded of a bumper sticker that would be quite familiar to anyone who lived in the UK during the late '70s and for many years afterward:

A dog is for life, not just for Christmas[26]

Agreement is not just for the kickoff meeting! Most of the changes that will be catalyzed by Kanban will stick only through agreement, so start as you mean to go on. Part III will help you with this; it describes a highly repeatable approach designed to generate *agreement in practice,* not just the slippery *agreement in principle* that falls away as soon as there is work to be done.

Change Management

Agreement is something that binds people together; it is a vital social skill, and it is a process. In an organizational context, agreement is useful shorthand for a valuable capability, *change management.*

The very idea of change management evokes a range of emotions, sometimes all at once. Done well, we hardly notice it—we feel part of the decision, not caring where it came from. Done badly, we resent it and we resist it.

These feelings seem to depend as much on the way the change is implemented as they do on the merits of the change itself. We find no contradiction in fully accepting the need for change and yet feeling personally insulted at the prospect.

And there are times when this resentment is fully justified. We see right through attempts to accelerate change through some inauthentic sense of crisis, invoking images such as the *burning platform,* when it is clear that neither lives nor livelihoods are under immediate threat. Ultimatums (involving some kind of threat) and edicts (no attempt at agreement at all, just one-way communication) hardly engender goodwill. Or we're left disoriented when management communicates changes of direction so inadequately or ineptly that no one really knows what is going on.

Three models for change management highlight the role of agreement in the change process:

1. Change led by a *change agent*
2. Change led by a change agent in combination with a *mentor*
3. Change led by the *change team*

---

26. Clarissa Baldwin. 1978. The Dogs Trust. http://www.dogstrust.org.uk/az/a/adogisforlife/

The Change Agent

Quite simply, change agents are people who cause change to happen. It's a broad term: They may be following some kind of implementation plan or they might catalyze change by communicating vision, leading by example, and so on. They may be the originators and instigators of change or delegated to carry it through.

Change agents often position themselves at the nexus of four sets of stakeholders:

1. The people whose work patterns will be most directly impacted by the change, often the change agent's primary focus
2. Those on the margins whose cooperation is likely to make the implementation of change easier
3. The customers who stand to benefit directly or indirectly from the change
4. Those to whom the change agent is personally accountable—line managers, a project board, perhaps shareholders

Most of us can recall some truly excellent and honorable examples of this model at work. Chapter 4 of the "blue book" describes how Dragos Dumitriu (now a friend and colleague of mine) sought to understand the problems of Microsoft's XIT team from the different perspectives of its team members, its customers, and the wider organization. Armed with an idea for a way forward, Dragos then engaged in some skilled "shuttle diplomacy" that built sufficient trust for a radical experiment. Shrewdly, Dragos avoided the big decision meeting until success was all but assured. Previously described as doomed to extinction, the XIT team progressed (in the words of the chapter title) "from worst to best in five quarters."

For every good story, I suspect there are numerous others that few would celebrate. Stories in which

1. The change agent believes that his or her main challenge is to *overcome the resistance to change* of team members.
2. Scant regard is paid to peers—they are "kept informed" rather than engaged.
3. The customer barely gets a mention.
4. Management gets good news right up until the moment when the initiative is abandoned.

These examples represent four stakeholder groups, each one poorly served. Let's see what a good mentor could bring to situations like these.

The Mentored Change Agent

At its most generic, the role of a mentor is to help someone see things as they really are and, through that process, to help guide the mentee (the change agent) toward better decisions. The mentor's job isn't, however, to deeply analyze the current situation through the mentee's eyes; it's more a process of drawing on relevant past experience with authenticity and authority and making sure that the mentee's own thinking undergoes a sufficiently thorough self-examination ("guiding by analogy," if you like).

In the XIT story, Dragos's mentor was David Anderson. They both worked at Microsoft at the time but were not close colleagues. Rather, it was David's published work that prompted Dragos to seek him out. David had no direct stake in Dragos's situation (they barely knew each other), but was able to provide Dragos with a new way to look at the problem.

In the Lean model (Chapter 14; see also Chapter 17), the mentor tends to be more involved, not necessarily independent of the situation. In fact, the lines between mentor, manager, and coach can seem rather blurred here. This may challenge your ideas of what these roles mean, but it seems to work, and it works because the focus of the relationship is on the situation and its problems, not on the mentee.

Grouping some of the challenges of change management by stakeholder group, here are some questions with which a mentor might gently guide the process:

1. **What solutions to the problem has the team considered?** This is a helpfully awkward question designed to find out whether the team is properly involved, and it can lead to a better framing of the problem if the change agent is too focused on a singular solution.

2. **What, specifically, have your peers (and others impacted) agreed to do?** It's not enough merely to inform; here, the mentor is guiding the mentee toward agreements that will stick. This is an opportunity to find wider benefits that otherwise would not have been considered.

3. **In what ways does the customer benefit?** A change that does not bring customer benefit is unlikely to qualify as an improvement.

4. **How does this look from the organization's perspective?** Is it cost effective? Does it align with strategy, policy, and values?

And there's a fifth consideration:

5. **Are we solving the right problem?** Again, this is about framing, but it is also about scope. Make the scope too narrow, and opportunity will be missed; but make it too broad, the mentee will have little hope of achieving meaningful success.

The Change Team

So many stakeholders! We have the instigators of change, those designing the details of the change, those implementing it, those impacted by it, and those benefiting from it. Then add the mentors, coaches, and managers of all the above. What if they were, by and large, the same people—a team? Wouldn't that be easier?

These things don't happen by accident, of course, and my Austrian friends and Kanban experts Klaus Leopold and Sigi Kaltenecker teach the strategies and skills that go with forming, participating in, and leading such *change teams*. I won't reiterate their work here (I encourage you to read their book or take one of their classes[27]), but pause for a moment to think about what happens when you bring all of those people in front of a kanban board.

The flow of work is visible, along with its hindrances. If there's a bigger problem to solve, the impact on the design of the process—which is right there in front of everyone—can be worked through. When there are tradeoffs to be explored, they can be made real by referring to actual work items.

What the change team is doing here is little different from what happens every day in front of the board. And this, of course, is the point: Kanban is integrating a change capability into the way the organization actually operates.

Any team that treats Kanban merely as a productivity tool misses this crucial point. Knowing this, you might still decide to keep it "under the radar" in the hope that an unstoppable viral spread will take hold, but at

---

27. As I write, an English translation of their German-language book, *Kanban in der IT,* is in the works, tentatively titled *Kanban in IT.*

what point do you invite the organization to make conscious decisions about direction, speed and objectives? That approach makes me distinctly uncomfortable.

The Kanban Method is built on **agreement**. Early agreement on the fundamentals sets the context and tone for everything that follows. At the next level of implementation detail, agreement represents skills that the organization can choose to develop. For the individual leader, agreement is a key discipline. Ad hoc implementations risk missing out on all of these.

❖ CHAPTER 9 ❖

Respect

Our ninth and final value is the third of our three *leadership disciplines*. What if all change could be conducted with **understanding, agreement,** and **respect**?

Respect underpins Kanban's third Foundational Principle:

FP3: Initially, respect current processes, roles, responsibilities, and job titles.

It is sometimes suggested that this is redundant. Why would the humane, *start with what you do now* method need to state this principle explicitly?

I can think of at least three important reasons for its inclusion:

1. It contains a pragmatic piece of change management advice: Don't start your initiative by looking at roles.

2. It points to an underlying philosophy: Find ways around obstacles to change.

3. No other principle or practice expresses "*respect for people*" quite so directly.

Don't Start with Roles

Changing the emphasis a little, let's summarize some earlier advice:

◆ Start with what you *do*, looking at how *what you do* meets (and fails to meet) the needs of people inside and outside the system.

◆ Organize work; let people self-organize around it, allowing the system to change as a result.

A premature focus on roles would fatally undermine this process. If they need to be defined formally at all, roles can be aligned to process once it has been established—in context—that the process is sufficiently fit for purpose. Meanwhile, we'll be making it very much easier for people to find that alignment for themselves. Why force the issue?

"Be Like Water"

A premature focus on roles has a highly undesirable effect: It provokes resistance. Regardless of the logic, it is tough to hear from someone else that different roles are required. No matter how positive the intent or how carefully it is explained, the message heard is one of inadequacy or irrelevance. No one wants to hear that they aren't meeting the needs of the organization; defensiveness is almost inevitable.

Kanban's folklore borrows the phrase, "*be like water*," from a rather unlikely source. Bruce Lee was famous for his martial arts, but rather less well known for his philosophy. He used that phrase more than once; we're referring to this particular expansion:

> Be like water making its way through cracks. Do not be assertive, but adjust to the object, and you shall find a way round or through it.[28]

The Kanban Method comes with the warning that roles are a rock, causing us sometimes to question the role-centric approach of other frameworks. Why start by insisting or implying that certain roles (product managers, project managers, business analysts, testers, and so on) are or aren't needed and appreciated? When there is so much to be done, why confront that rock when you can so easily go around it?

There will be other rocks, of course, and whether you confront them or find a way around them is a matter of judgment. All else being equal, we recommend you choose the path of adaptability, the one that keeps the process of change on the move. Another very good guide is **respect**: Are you confronting this rock just because the plan has it next on the list, or because people's needs can't properly be met until this issue is dealt with?

---

28. I have not been able to establish definitively the origin of this longer quote. Whether it was Bruce Lee himself or one of his many followers or commentators that first wrote it, I cannot reliably say.

Respect for People

I am very happy to acknowledge here a debt to Toyota and Lean. Several of Kanban's values correspond to Lean principles, but few more so than **respect**. *Respect for people* has long been identified as a so-called "pillar" of Lean. Along with *continuous improvement*, it is now (since 2001) one of only two pillars of the Toyota Way, the Toyota company's stated management philosophy.

However, the objective of this chapter is not to explain what *respect for people* means inside Toyota or for practitioners of Lean methods; that is covered in Chapter 14. Instead, this chapter aims to do the following:

1. Show how the Kanban Method measures up to the test that **respect** represents

2. Show how **respect** can be a helpful guide when implementing Kanban

As we approach the end of this values-based introduction to the Kanban Method, it's a good time to recap, value by value.

Transparency

With **transparency**, Kanban encourages us to visualize, to make policies explicit, and to implement feedback loops. These practices are respectful in intent if they are done with the expectation that people will make better and more meaningful choices when given the information, the tools, and the opportunity.

Respect is a two-way street. Don't forget that the organization must live with the outcomes of those choices; perhaps its leaders should earn some respect for creating the conditions in which those choices can be made.

There is respect inherent in some of transparency's more advanced concepts, too. Embracing variety—rejecting, for example, the idea that every piece of work must be assigned a date—isn't just "nice," it improves outcomes overall and it enhances the system's resilience. Resilience is further enhanced when people have the space to act with meaningful autonomy, and systems can self-organize. Who wants to work inside a fragile system? Not me, that's for sure.

Is there such a thing as "too much transparency"? Some of my more radical friends argue quite forcefully that transparency should be unlimited.

Even if they are right (and there are some exciting examples out there of radically transparent companies) I would ask this: Is it possible to implement transparency thoughtlessly or recklessly? I'm sure that it is.

Balance

Kanban's first and most direct objective of **balance** is to match demand and capability. Its most basic tool is the work-in-progress limit, which helps people by addressing both overburdening and starvation and by making their impediments more visible. That seems respectful; but again, respect can act as a check. Rushing to impose aggressive limits without considering individual capability or preference can have negative consequences.

Balance extends respect to other stakeholder groups. Risks and benefits can be apportioned equitably. Respect should cause us to reject "zero-sum improvements" that benefit one set of stakeholders at the expense of another.

Collaboration

These aspects of **collaboration** seem very well grounded in respect for people:

◆ Working collaboratively to deliver something of value

◆ Improving the system, collaboratively solving deep-rooted problems

◆ Treating collaboration as an end as well as a means, creating greater opportunity for interactions of higher quality

Of course, not everyone wants to work collaboratively one hundred percent of the time. People need space, too! Again, respect is a check.

Customer Focus

In the search for meaning in our work, "respect for the customer" isn't a bad place to start. There's gratification to be found in meeting needs, more still in anticipating them; I have no doubt that **customer focus** is a humane value. Will our anticipatory capability develop substantially in the absence of respectful conversations and self-examination? Unlikely.

Flow

To value **flow** is to place sufficiently high value on smoothness and time-liness that it sustains improvements to the system. Flow can be measured, but it shouldn't be reduced only to numbers—it needs to be experienced and enjoyed too.

Once you learn to appreciate flow, its absence hurts. Sometimes, how-ever, an abrupt change of direction is justified; on these occasions we might say "*value trumps flow*"[29] and accept some short-term discomfort. If we're too accepting of this, though, our commitment to flow and its benefits—both economic and humane—may be called into question.

Leadership

Leadership at every level is clearly about people, and not just the person who happens to occupy the top position or the role of change agent. Kanban's approach to **leadership** involves seeing the organization's poten-tial as a self-sustaining system, modeling behaviors we want to see per-petuated, taking opportunities and creating them for others, and taking appropriate risks. This is how systems change from the inside.

To restrain leadership's excesses, we have the *leadership disciplines* of **understanding, agreement,** and **respect.** And if you still struggle with the value of leadership, reflect instead on *opportunity.*

Understanding

Understanding is a call to a rich appreciation of the system that includes not just functions and activities but the people inside and outside the sys-tem, their capabilities, needs, and frustrations, too. *Start with what you do now* is also about safety and *survivability*, and while we would not argue that organizations should survive at all costs, neither would we wish unnecessary harm on those who depend on them.

To lead with understanding we must seek to avoid *complacency, bravado, and tampering.* Respect for personal and corporate safety is a good guide.

---

29. This is the first half of what David calls the *Lean Decision Filter*: "Value trumps flow and flow trumps waste elimination." His point is that waste elimination is third in priority, contrary to some popular characterizations of Lean. Value and flow aren't usually in opposition.

Agreement

In **agreement**, we make a specific and long-term commitment to the pursuit of evolutionary change, committing indirectly to developing the organization's capability for change. This investment in adaptability is the second leg of a survival strategy, and it, too, is an investment in people.

Sponsors of change (any and every change, not just the big initiatives) have a special role to play. They can model disciplined leadership by expecting it to be conducted respectfully, with understanding and agreement. Not that agreement should imply a boring, consensus-driven monoculture; evolutionary change needs creativity, which in turn requires that diversity be respected, embraced—even celebrated.

The Humane, *Start with What You Do Now* Method

Depicting Kanban as "the humane, *start with what you do now* method" highlights what I believe are two of the Kanban Method's most differentiating features. They're more differentiating than even the sticky notes and WIP limits, and that's saying something!

Kanban is humane, not just because it has noble ideals, not just because it leads to better conditions for people, but because it is designed to be implemented in a way that is consistently respectful. Noble aims don't justify unsafe organizational transformations. The goal of better conditions in the longer run doesn't excuse riding roughshod over people in the interim. To my knowledge, few methods can claim Kanban's internal integrity in its approach to implementation.

Also, there is something uniquely respectful in *start with what you do now*. It respects the journeys of all the people—past and present—who have brought the organization to where it is now. For all the good and the bad, it respects context, it celebrates survival, and it makes it easier to assume—however naively optimistic an assumption this might be—that people mostly did their best. I like that.

❖ CHAPTER 10 ❖

Patterns and Agendas

Starting with the most concrete and familiar, finishing with the most abstract and philosophical, I've taken you through the values one by one, introducing you to the Kanban Method. I hope that along the way I've given you some sense, too, of what it feels like to practice it for real.

I would be the first to admit that it takes effort to remember nine values, and I discovered very quickly that a little structure can be very helpful. You've seen hints of this already: the reflections after Chapters 3 and 6, for example, and the label of *leadership disciplines* applied to the final three values.

I found that I looked at groups of values differently when I named them. These names have proved to be rather less stable than the values, however, and it took some time to finalize them. Eventually, in an intense collaborative effort that spanned two conferences over six days and involved myself, David Anderson, Markus Andrezak, Andy Carmichael, and Patrick Steyeart, we finally settled on three names that collectively describe Kanban's *agendas for change*.

Before we look at those, I give an honorable mention to Steve Tendon, one of the first people to take the nine values and do with them something I really hadn't expected.

"Noble Patterns"

Steve's interests lie at the intersection of Kanban and two other important bodies of knowledge: the *Theory of Constraints* (see Chapter 12) and *Scrum* (Chapter 13).

In the book *Tame the Flow* (co-authored with Wolfram Müller), Steve aligns the values with three "patterns" as follows:

1. Community of Trust: **understanding, agreement, respect, transparency**, and **collaboration**
2. Unity of Purpose: **flow, customer focus**, and **balance**
3. **Leadership** (on its own)

Community of Trust and *Unity of Purpose* are Steve's "noble patterns," highly desirable, only rarely seen fully realized together, and associated with the most productive environments. Not quite so noble, it seems, is **leadership;** perhaps he sees it as too common (if only that were true!) or perhaps it is exercised poorly too often.

Steve found different groupings from mine and his names for them are vivid. His patterns preceded my work on values, but the alignments make good sense and the overall effect is pleasing and memorable. I love how they subtly evoke some extra meaning in the values. Very gratifying!

Agendas for Change

Does Kanban in some neutral way just create the conditions for change, or does it come with its own biases? Do the method, its practitioners, and their host organizations need direction—in the form, perhaps, of an external *true north* (Chapter 14)—or will they steer themselves? As a community, we've considered these questions several times.

By now you've seen enough of Kanban that you won't be surprised to find that the answers turn out to be rather inclusive:

◆ Yes, Kanban creates the conditions for change—in fact, developing the organization's capability for change is one of its main objectives.

◆ Yes, whether expressed as a "true north" or an evolutionary "fitness function," it's important to have measures of success.

◆ Yes, not only does Kanban have some built-in biases, it's helpful to make them explicit.

There is one "no" implied by that list of yeses: No, Kanban is not neutral. Today this seems obvious: How could something so intertwined with values—where, for the most part, "more is better"—ever be described as neutral?

Enter Kanban's three *agendas for change*. You can think of these as adoption approaches, which you can choose and communicate according to your organization's needs, appetites, and ambitions:

◆ **The sustainability agenda**: This describes a common approach to Kanban adoption at the individual and team levels. Often the motivation is the relief that **transparency, balance,** and **collaboration** can bring from unsustainable practices and workloads.

◆ **The service orientation agenda**: This adds **customer focus, flow,** and **leadership**; with these come a significantly more outward-looking approach to engagement and adoption. Typically, the motivation here is the desire for significant improvements to customer delivery.

◆ **The survivability agenda**: This is the most overtly cultural agenda of the three. With the *leadership disciplines* of **understanding, agreement,** and **respect** come meaningful personal and organizational commitments to pursue fitness through adaptability and the capability to change.

The Sustainability Agenda

When individuals and teams are struggling under the sheer weight of unfinished work, introducing kanban systems can bring swift relief. Improvements in predictability help win trust, from which comes the space to keep exploring opportunities for further change.

Where *what you do now* is well defined, finding alignment to the values of **transparency, balance,** and **collaboration** tends to be straightforward and uncontroversial. Where it is not, Kanban presents a low-risk opportunity to better understand it.

With its echoes of the Agile Manifesto's "sustainable pace," the sustainability agenda often resonates strongly with teams that would identify themselves as Agile (see Chapter 13). Moreover, many of their existing practices and artifacts can be interpreted in terms of transparency, balance, or collaboration, the last of these being a key theme in the Agile Manifesto (itself a statement of values).

Some caution is in order: Although transparency and balance scale easily (right up to the level of corporate initiatives and themes), the same cannot be said of collaboration. On its own, this agenda yields benefits to external stake-holders somewhat indirectly and its focus can easily turn inward. Without an external "pull" to improve, progress tends to slow, perhaps even stall.

The Service Orientation Agenda

If the sustainability agenda is best described as a practice-based approach, the service orientation agenda is based on engagement. You won't get far with **customer focus**, **flow**, and **leadership** on your own!

Progress in these comes from:

♦ Looking critically at what you do now using the tools presented in Chapters 4 and 5

♦ Improving coordination end-to-end, preventing WIP from accu-mulating (or worse, getting mislaid) between service boundaries

♦ Shaping demand upstream

♦ Gaining a better understanding of the gap between customer expectations and the capabilities of the system, and putting in feedback loops to control it

♦ Tracking changes, making sure that appropriate credit is given for improvements large and small

The service orientation agenda is a natural choice wherever there are clear external drivers for change. With these comes sponsorship, easing the implementation of the bigger feedback loops.

The Kanban Lens

David Anderson helpfully summarizes how the service orientation agenda can be introduced with another 2013 innovation, the *Kanban Lens*. This describes how to view *what you do now* as a set of services that can be improved.

Recognize that:

1. **Creative knowledge work is service oriented**—much of its value is in the needs that it meets for other people, directly or indirectly.

2. **Service delivery involves workflow**—activities depend on other activities; states follow states; work starts and finishes outside the system.

3. **Workflow involves a series of knowledge discovery activities**—indeed, this is a defining characteristic of creative knowledge work.

Then:

1. **Map the knowledge discovery workflow**—capture a high-level understanding of the sequence of key activities and work item states necessary for completion.
2. **Pay attention to how and why work arrives**—understand the motivation, mechanisms, and patterns of arrival.
3. **Track work as it flows across and between services**—visualize its progression through to completion and validation.

Compared with the sustainability agenda which scales mainly vertically, the services orientation explicitly scales Kanban horizontally across services, both end-to-end and between dependent services.

The Survivability Agenda

When it comes to the survivability agenda, "choice" seems too weak a word—it's a strategic decision, an imperative. Here, the aim is to institutionalize evolutionary change so that a more adaptable and resilient organization can emerge as it pursues fitness for purpose.

In contrast to the bottom-up and middle-out approaches of the first two agendas, the survivability agenda starts with senior-level commitment to the disciplines of **understanding, agreement** and **respect**; the rest follows, with sponsorship, scope, and objectives assured.

Your Organization's Agenda

However we might choose to organize them, Kanban's values are ultimately about people. People with transparency over and balance in their workloads. People whose relational needs—opportunities for collaboration, leadership, and so on—are recognized. People able to participate in a flow of work that meets meaningful customer needs. People who can work in the knowledge that although change might be a constant presence, it is safe and it is purposeful.

If your organization would struggle to identify with those aims even for the long term, you might be forced to accept that Kanban might not be a good fit. For most organizations, though, it's more a question of priorities. Which values best capture your organization's way forward from where it is now? How would you describe them collectively? That topic is addressed toward the end of Part III.

How Did We Do?

My 18-month Budapest experience saw two IT teams adopt Kanban, the second a few months after the first. Using the new language of values and agendas as I look back on the events of 2009–10, here's a retrospective:

- ◆ The sustainability agenda felt natural internally, and there is strong evidence that other parts of the company came to understand what we were doing. The first team settled into its kanban system very quickly and continued to refine it. The second adoption was even more rapid, everyone benefiting from the example of the first. It then became clear that we had to get the project portfolio under control; the degree to which the portfolio WIP reductions were offered up by project sponsors seems remarkable to me even now.

- ◆ The service orientation agenda does a good job of representing the direction and drive of our implementation. Chapter 4 explains how we renewed our **customer focus**. As described in Chapter 5, we were always attentive to **flow**, and over time our capability was transformed. Opportunities for **leadership** (Chapter 6) were plentiful, some by design, others created by my unusual, alternating-weeks onsite/offsite work pattern.

- ◆ Our relationship with the survivability agenda was more implicit than explicit and more internal than external:
  - ◆ It is fair to say that **transparency** preceded **understanding**. For both teams, our top reason for introducing Kanban was need for a better grasp of where we were. I had sufficient experience as a manager to have some idea of what we were getting ourselves into organizationally, but I can't claim that we started with a deliberate effort to achieve a level of shared understanding or to take in multiple external perspectives in a very structured way.
  - ◆ We practiced **agreement** within the team, but only belatedly did I come to realize that we had relied too much on the external

sponsorship from my managing director and I had not worked hard enough on agreement with my peers in the management team. This slowed some decision making and delayed some key deliveries.

+ No issue with **respect**—that was a given.

Within the bubble of my formal span of control, we were, in the end, successful in following all three agendas. If I had the same opportunity again and were to go in explicitly with a service orientation agenda, I would expect similar results (projects delivered, lead times shortened, a more focused project portfolio, and a sustainable process), but sooner.

It is more than a little chastening that the survivability agenda captures our areas for improvement so well. We needed **agreement**, and we might have obtained it sooner had we done more to create shared **understanding** of where we were.

There's a survivability-related postscript to this story. My main objective was all about securing the longevity of the organization, though not necessarily in its existing form. My task was to get the IT function, its key applications, and its infrastructure into good enough shape that they wouldn't represent risks to any potential future merger. I left the company on good terms after this scenario materialized and I helped take them through a merger with a larger competitor. It soon looked like a reverse takeover: Mark Dickinson, my former managing director, became CEO of the merged company not long after. Job done!

Five Years On

Fast-forward to the present, and we now know that our experience is highly repeatable. Teams that implement Kanban feel the benefits quickly. Where attention is paid to end-to-end flow and to anticipating customer needs, dramatic improvements in service delivery are possible. Only a few companies seem to make an explicit commitment to the pursuit of adaptability and fitness for purpose, but some exciting examples do exist outside David's book.

Meanwhile, the Kanban community has gained its own identity and its knowledge base has grown and matured. Not, however, that it claims (or even seeks to claim) that it has everything covered. Instead, we look to *models*, some celebrated, some less well known. They're the subject of Part II.

❖ PART II ❖

Models

Part I described how the Kanban Method works, using the values as a means to get inside the rather particular way Kanban practitioners look at organizational challenges. There were some passing references to outside influences, but I didn't want those to obscure that main theme.

In Part II, the values take on a supporting role. Center stage goes to *models*, bodies of knowledge I wish to celebrate, both for their historical relationships to the Kanban Method and its community, and for what they have achieved in their own right. The task of the values here is to give us a basis for comparison and integration.

Recall the Kanban Method's sixth Core Practice:

CP6: Improve collaboratively, evolve experimentally (using models and the scientific method).

There is something delightfully flexible about "*using models*" here. It can mean

◆ Replicating the example of others

◆ Following some set pattern, template, or routine that gives structure to one's actions or thinking

◆ Understanding the world based on a defined set of assumptions

◆ Expecting certain kinds of outcomes as consequences of those assumptions

When the model in question is a body of knowledge of the scale of Systems Thinking and its related disciplines, Theory of Constraints, Lean, or Agile, using models" can mean all of these things at the same time. These are presented briefly in Chapters 11 through 14.

Each of these models represents an important paradigm shift:

♦ Systems Thinking is a move away from reductionism toward a holistic view of systems.

♦ Theory of Constraints replaces the traditional cost focus of improvement with one more properly aligned to system goals.

♦ Lean seeks efficiency through the pursuit of flow rather than the maximal utilization of resources.

♦ Agile replaces planned, project-driven delivery with an evolutionary and collaborative approach.

This book won't turn the novice reader into an expert in any of these models (there are other books for that); but to fully grasp the intent of the Kanban Method, these paradigm shifts need to be appreciated.

Chapter 15 covers some economic concepts essential for effectively managing **flow**.

Chapter 16 gathers into one place the principles, practices, and other core concepts related to the Kanban Method. It also covers some of its specialized applications, including Personal Kanban, Portfolio Kanban, and Scrumban.

Finally, Chapter 17 covers some smaller models. Some of these support a deeper understanding of earlier parts of the book; others provide some useful tools for Part III.

❖ CHAPTER 11 ❖

Systems Thinking, Complexity, and the Learning Organization

Systems Thinking[30] together with the related science of *complexity* covers a wide range of disciplines. This very short excursion through these fields starts with *leverage points* and the Systems Thinking methodology of *System Dynamics*. Via *nonlinearity* and *unpredictability* we visit *complexity, causality,* the *Cynefin framework,* and *complex adaptive systems.* The end of our tour sees these integrated with the concept of the *learning organization.*

At the end of this chapter we show how these concepts have influenced the design of the Kanban Method and how they can help us apply it more effectively.

Systems Thinking

Broadly, Systems Thinking is concerned with understanding how systems behave as a whole. In contrast with *reductionist* or *analytical* approaches that work to understand system components in isolation, Systems Thinking is more holistic, emphasizing the relationships, interactions, and influences among components and the behaviors and outcomes that emerge from them.

---

30. I treat Systems Thinking, Agile, and Lean as proper nouns when referring to them as bodies of knowledge, approaches, or communities. I try not to use them as adjectives.

Systems Thinking has been applied to social, economic, and political systems, sometimes to systems that are very large indeed. A well-known example is the landmark study *The Limits to Growth*[31] by Donella H. Meadows et al., which explores the relationships between economic growth and finite resources such as minerals and non-renewable energy.

At a rather more familiar scale, John Seddon has taken Systems Thinking into the design of service operations in the public sector. Gerald M. Weinberg, noted writer on the psychology and anthropology of software development, uses Systems Thinking to explain a number of phenomena we regularly see inside our organizations.

Other influential *systems thinkers* include

◆ W. Edwards Deming, the pioneer of *quality management* who played a key role in rebuilding Japanese industry after World War II

◆ Peter Drucker, who—as far back as 1959—coined the term *knowledge worker*, whose cause this book seeks to serve

◆ Multiple generations of the Toyoda family, together with Taiichi Ohno, the father of the *Toyota Production System* (TPS), from which *Lean* is derived (see Chapter 13)

Two things seem to separate systems thinkers from the rest:

1. The discipline to consider the bigger system—the system of concern taken together with some surrounding context

2. The ability to identify the hidden influences between system components and between the system and its environment

These are good skills to learn. The better we understand our system's relationships with its environment, the more likely it is that we will be able to identify and implement effective interventions.

Leverage Points and System Dynamics

Systems Thinking helps us identify *leverage points*, places in the system where intervention will be disproportionately effective. With luck, these will lie within our span of control or (failing that) our sphere of influence. All is not lost even when they don't; Systems Thinking may help us find *alignment* in the form of interventions that deliver benefit to some bigger whole.

---

31. Meadows, Donella H., Dennis L. Meadows, Jørgen Randers, and William W. Behrens III. 1972. *The Limits to Growth*. New York: Universe Books.

Meadows is responsible for one of my favorite pieces of Systems Thinking, her *twelve leverage points to intervene in a system*. Strikingly, they are numbered from 12 down to 1, with the least effective leverage points first:

12. **Numbers**: Constants and parameters such as subsidies, taxes, and standards

11. **Buffers**: The sizes of stabilizing stocks relative to their flows

10. **Stock-and-flow structures**: Physical systems and their nodes of intersection

9. **Delays**: The lengths of time relative to the rates of system changes

8. **Balancing feedback loops**: The strength of feedbacks relative to the impacts they are trying to correct

7. **Reinforcing feedback loops**: The strength of the gain of driving loops

6. **Information flows**: The structure of who does and does not have access to information

5. **Rules**: Incentives, punishments, and constraints

4. **Self-organization**: The power to add, change, or evolve system structure

3. **Goals**: The purpose of the system

2. **Paradigms**: The mind-set out of which the system—its goals, structure, rules, delays, and parameters—arises

1. **Transcending paradigms**

Some of the terminology of interventions 11, 10, 8, 7, and 6 may need further explanation. Meadows was a practitioner of *System Dynamics*, a methodology that originated in the domain of industrial design. *Stock* therefore initially represented a quantity of physical material, and *flow* referred to the transfer of those materials from one place or activity to another. As the application of System Dynamics broadened, stocks came to refer to other kinds of tangible things, such as animal populations or money, and to intangible things, such as quality or trust. The more abstract the stock, the more subtle the flow.

Complexity

Systems that contain delayed signals and feedback loops can exhibit behavior that is very hard to predict, even when they remain fully *deterministic*, completely immune to randomness. These systems are *nonlinear*—they can't be explained simply as the sum of their parts. Some surprisingly simple systems are so sensitive to initial conditions, to measurement error, or to the tiniest perturbations (the *butterfly effect*) that their behavior appears chaotic.

Business systems can suffer from this too. A classic example is the *bullwhip effect*, sometimes known as the *Forrester effect*, named after Jay Forrester, one of the founders of System Dynamics. This effect explains violent swings observed in supply chain inventories. Suppliers lack perfect information and might assume that short-term changes in demand have long-term significance. This drives them simultaneously to cover the immediate demand and to increase their buffer stock, causing an amplified signal to be sent up the chain. The next supplier in the chain further amplifies the signal. Given a long enough chain, stocks might fluctuate wildly.[32]

Another way to magnify the effect of amplification is via feedback loops. Unrecognized or poorly designed feedback loops can be devastating to a system.

Weinberg describes the common vicious circle shown in Figure 11.1.[33]

In a pessimistic reading of that picture, a decline in understanding the value of quality reduces the motivation to achieve quality. This in turn reduces the understanding of how to achieve quality, thereby decreasing the motivation to measure its value, and so the cycle continues ever downward.

Fortunately, there is an optimistic reading of this picture. Given the right initial conditions or a sufficiently strong upward intervention to the otherwise downward-spiraling system, this *reinforcing loop* can behave as a virtuous circle. The understanding provides the motivation, the achievement reinforces the understanding, and so it continues.

These systems are simple enough to understand, but implications for the manager are profound. The behavior of the system as a whole cannot be pinned down to any one of its parts. It is difficult to separate cause from effect. Understanding the system will certainly help the manager find

---

32. The *beer game* is a simulation that demonstrates this effect.

33. Weinberg, Gerald M. 1992. *Quality Software Management, Volume 1: Systems Thinking*. New York: Dorset House.

Figure 11.1   Vicious or virtuous? A feedback loop in quality

helpful interventions, but it would be a brave manager indeed who would state with certainty exactly how they will play out. Managers don't determine how the system behaves; they merely interact with it.

Causality and the Cynefin Framework

Building on these ideas, Dave Snowden describes in the *Cynefin framework*[34] five domains that characterize *causality* within systems:

♦ **Obvious** (or in older literature, **Simple**)—where we can easily categorize what we see and respond accordingly, perhaps with a *best practice,* whose outcome will be readily apparent

♦ **Complicated**—unobvious to a casual observer, but where an expert could understand what is going on sufficiently well to select an appropriate *good practice* with a predictable outcome

♦ **Complex**—the domain of *emergent behavior,* only in retrospect can we fully understand the impact of our interventions

---

34.   Snowden, David J. and Mary E. Boone. *A Leader's Framework for Decision Making,* Harvard Business Review, November 2007.

- ◆ **Chaos**—where causal relationships disappear entirely, even in retrospect
- ◆ **Disorder**—the state of not knowing which of the other four domains applies

These domains are distinct. In particular, the *complex domain* isn't just a "more complicated" version of the *complicated domain*. The challenge of the complex domain—the domain that most interesting business systems inhabit—lies in nonlinearity. We can't find better experts who are more capable of predicting accurate outcomes. Strategies that seem to work in the simple and complicated domains either don't work at all or they seem to work for a while and then break down.

Cynefin offers some reassurance in its description of the complex domain: We can hope to make sense of things retrospectively. In other words, we can look at the way things are (or have been) and understand them as the product of previous decisions or conditions, internal and external.

Turning this around, we will soon start to see the result of current interventions. We can plan to validate at some future time whether they have been beneficial, harmful, or of limited impact. With some added forethought, we might predict several of those potential impacts in advance, readying ourselves to actively *amplify* the beneficial ones and *dampen* the harmful ones. These *safe-to-fail experiments* enhance *PDCA* (Chapter 3) for those cases where intervention might easily do harm or where hoped-for emergent benefits may need early recognition and nurturing.

Cynefin also allows us to see some interventions as moving aspects of the system from one domain to another—pulling them back from chaos (or carefully dipping into it), simplifying the complicated, and so on. The specific practices of other frameworks (Agile methods, for example) can usefully be understood in this light.

Complex Adaptive Systems

Designing for complexity doesn't just mean designing better experiments. If unpredictability is a constant presence, we need to design for *resilience* and *adaptability*. Complexity science offers a model: *complex adaptive systems (CAS)*, multiple levels of loosely coupled self-organizing systems.[35]

---

35. Note that Meadows identified self-organization as one of the most effective intervention points— it's at number four on her list.

The adaptive nature of CAS is vital for systems that live in competitive environments in which only the *fittest* will thrive. Adaptive systems are more likely to keep finding configurations that maintain their edge. Even away from the front line, adaptive systems with well-designed feedback loops can still seek out increasing *fitness*. These are strategies for *evolutionary change*.

Knowledge, Learning, and the Learning Organization

What if the problem isn't so much the current shape of the organization but the thinking that gave rise to it? Of course these aren't so easily separated. As Weinberg observes:

When the thinking changes, the organization changes, and vice versa.[36]

It follows that adaptability (the ease by which a process or a practice can change) goes hand-in-hand with an appetite for learning (which changes thinking). This is good news: Work on either side of this equation and there will be impact on the other. Better still, work on both together.

We finish this section with one particular vision of the Learning Organization, namely, Peter Senge's. Before that, we need a couple more pieces of the puzzle. We look at the knowledge needed by managers as described by W. Edwards Deming, then at a simple but powerful model of organizational learning described by Chris Argyris.

The Deming System of Profound Knowledge

At the start of this chapter I introduced W. Edwards Deming (1900–1993) as an influential Systems Thinker. In fact, it's hard to think of a Systems Thinker who's had a greater influence on the Kanban community, although Eli Goldratt (see Chapter 12) comes close.

Deming began with Systems Thinking as a foundation and then added three further elements to form his *System of Profound Knowledge (SoPK)* for managers in business:

36. Weinberg, Gerald M. 1992. *Quality Software Management, Volume 1: Systems Thinking*. New York: Dorset House.

1. **Appreciation of a system**: understanding business systems together with their context of suppliers and customers
2. **Knowledge of variation**: understanding the causes and impacts of variation in quality and the proper use of statistical methods
3. **Theory of knowledge**: challenging management to learn by systematically developing, testing, and applying theories
4. **Knowledge of psychology**: understanding people and what motivates them; also the psychology of change

Each of these is useful individually, but the whole is significantly greater, and it highlights significant weaknesses in prevailing management thinking. For example:

◆ It is wasteful (or worse) for managers to *tamper* with a system, addressing each random variation as though it had an *assignable cause*. Equally, one should not dismiss variation that may have specific causes in the system as though it was just the result of natural randomness.[37]

◆ Many businesses try to motivate people through monetary and other rewards that are driven as much by random variation, system behavior, and the current business climate as they are by personal contribution. It has long been known that these strategies can be highly damaging, but these practices persist because the system promotes the people that support them.

Deming expanded on his System of Profound Knowledge in *14 Points for Management*, reproduced in Appendix A. More than twenty years later, it would seem that there is much learning still to do!

Argyris and Double-Loop Learning

How do organizations learn? Behavioral psychologist and management thinker Chris Argyris is associated with a number of significant contributions on this topic, three of which in particular are well known inside the Kanban community:[38]

---

37. Russell Ackoff, another great systems thinker, might categorize the latter error as an "error of omission." He considered these the most insidious management errors of all.

38. We're grateful to Benjamin Mitchell for championing Argyris' work.

◆ The *Ladder of Inference,* which helps to explain and deal with the very different conclusions that colleagues can reach

◆ *Espoused theory* versus *theory-in-use,* a way to seek out learning opportunities by looking at the differences between what people and organizations profess versus what they actually do

◆ *Double-loop learning,* a simple model that very elegantly describes a process of deep learning

The first two of these are outside the scope of this book. Double-loop learning is, however, very relevant. It is diagrammed in Figure 11.2.

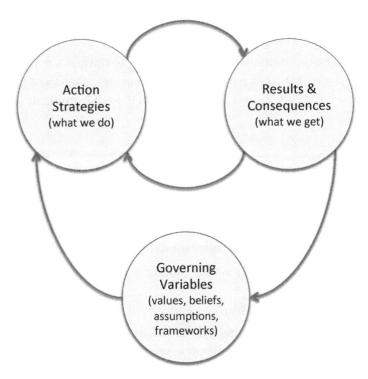

Figure 11.2   Double-loop learning

The inner loop in Figure 11.2 describes *single-loop learning.* Most learning is single-loop: We adjust our *action strategies*—our immediate goals, plans, and moves—according to the results we are observing. Single-loop learning comes easily, even unconsciously. Single-loop learning is very efficient when the goal is to keep a system under good control in a predictable environment.

But as we've seen, businesses don't exist in predictable environments. Even their internal systems can be deceptively unpredictable. Therefore, we can't expect organizations to remain competitive indefinitely if its people aren't engaging from time to time in a more rigorous kind of learning, namely, double-loop learning.

In double-loop learning, instead of just quickly recalibrating when results don't match expectations, the learner digs deeper. Assumptions are challenged. Mental models get discarded and replaced with new ones. Implicitly or explicitly, the change in thinking is reflected in new ways of working.

This unlearning and relearning is hard work! Realistically, no one has the energy to engage in double-loop learning all the time, but neither can an organization expect to remain viable if it remains complacent for long. Somehow, the conditions for frequent double-loop learning must be created.

The Learning Organization

As in Deming's *System of Profound Knowledge*, Peter Senge's model starts with Systems Thinking. Compared with Deming, however, Senge is less focused on the manager, more focused on individuals and teams. His five characteristics of a learning organization are as follows:

1. **Systems Thinking**: understanding and managing the organization as a whole
2. **Personal mastery**: individuals committing to the process of learning
3. **Mental models**: assumptions and theories made explicit, open to inquiry and challenge in an environment of trust (the conditions for *double-loop learning*)
4. **Shared vision**: the challenge that unifies and energizes the organization; higher purpose
5. **Team learning**: the social mechanisms that accelerate individual learning; the structures that build organizational learning from the learning of individuals

That last one is alluded to in the title of Senge's bestseller, *The Fifth Discipline.*

Senge has described the Learning Organization as:

> *. . . a group of people working together collectively to enhance their capacities to create results they really care about.*[39]

There is insufficient business purpose in that description for it to form a complete, unifying vision for most organizations, but as someone working in the field of organizational change, I find it compelling. When I see the organizations I'm working with move toward that objective, I'm seeing results that *I* really care about.

Systems Thinking and Kanban

The Method's Design

Several elements of the Kanban Method have their roots in the models outlined in this chapter. In particular:

◆ The **transparency** practices (Chapter 1) create new leverage points, making the system more open to challenge and improvement. Furthermore, they can promote self-organization, a strategy for resilience in the presence of uncertainty.

◆ Core practice 5, *Improve collaboratively, evolve experimentally,* connects **collaboration** (Chapter 3), knowledge, experimentation, shared learning, and evolutionary change.

◆ Kanban shares with Lean (Chapter 13) a Systems Thinking approach to **leadership** (Chapter 6). In the long run, organizations get the leadership they deserve, the kind that their system recruits, encourages, and promotes. It follows that the most enduring organizations are those that have paid attention to this.

◆ Albeit implicitly, the first foundational principle, *start with what you do now,* points both to Systems Thinking and to evolutionary change. I've added some extra emphasis by abstracting from this principle the **understanding** value (Chapter 7)—the goal is for

---

39. Fulmer, R. and J. B. Keys. "A conversation with Peter Senge: New developments in organizational learning," *Organizational Dynamics,* vol. 27, no. 2, p. 33 (Autumn 1998).

organizations to value understanding and to have the discipline to make it the precursor to change.

♦ Evolutionary change is explicit in the second foundational principle, *agree to pursue incremental, evolutionary change* (see Chapter 8, **agreement**).

Earlier chapters have made it clear that the Kanban Method leaves room for interpretation. This is a strength. It is articulated sufficiently clearly for a community to rally around it, yet it is applied with sufficient diversity that its community continues to learn, to develop lower-level practices, and to share experiences. It is satisfying to observe that the Kanban community itself demonstrates in some measure all five of Senge's characteristics of the Learning Organization.

Application

Neither Kanban nor Systems Thinking should be one-off exercises or specialist, "ivory tower" disciplines that are kept separate from where the "real work" is done. John Gall advises this (sometimes known as Gall's law):

A complex system that works is invariably found to have evolved from a simple system that worked. The inverse proposition also appears to be true: A complex system designed from scratch never works and cannot be made to work. You have to start over, beginning with a working simple system.[40]

Don't worry that every step must be right the first time. Keeping the system in motion with *safe-to-fail experiments*, there's a limit to how much harm any one step can do, and any local difficulties will soon shake out. Under these conditions, *suboptimization*—the localized improvement that makes things worse globally—is less a problem than the loss of momentum caused by fear of it. "The perfect is the enemy of the good," as they say.

With those evolutionary thoughts in mind, consider the following as exercises:

♦ Reflect on these *stocks*, identifying others of similar significance in your current context:
  ♦ Unmet need

---

40. Gall, John. 2003. *The Systems Bible: The Beginner's Guide to Systems Large and Small*, 3rd ed. Walker, MN: General Systemantics.

- Risk appetite
- Goodwill
- Quality
- Trust

What influences them? How do the nine values from Part I interact with them?

- Carefully review Meadows' leverage points. What parallels can you find with the Kanban Method? Which of her interventions might apply in your current situation? How might Kanban help?
- Try to map as many aspects of your current situation as you can to the domains of the Cynefin framework. How could you use the practices of Kanban to help move them from one domain to another?

Further Reading

Systems Thinking

Deming, W. Edwards. 2000. *The New Economics,* 2nd ed. Cambridge, MA: MIT Press.

Gall, John. 2003. *The Systems Bible: The Beginner's Guide to Systems Large and Small,* 3rd ed. Walker, MN: General Systemantics.

Meadows, Donella H. 2008. *Thinking in Systems: A Primer.* White River Junction, VT: Chelsea Green.

Seddon, John. 2003. *Freedom from Command and Control: A Better Way to Make the Work Work.* Buckingham, UK: Vanguard Consulting Ltd.

Weinberg, Gerald M. *Quality Software Management, Volume 1: Systems Thinking.* 1992. New York: Dorset House.

Learning

Kahneman, Daniel. 2011. *Thinking, Fast and Slow.* New York: Farrar, Straus and Giroux.

Noonan, William R. 2007. *Discussing the Undiscussable: A Guide to Overcoming Defensive Routines in the Workplace.* New York: Wiley/Jossey-Bass Business.

Patterson, Kerry, Joseph Grenny, Ron Mcmillan, and Al Switzler. 2011. *Crucial Conversations: Tools for Talking When Stakes Are High,* 2nd ed. New York: McGraw-Hill.

Patterson, Kerry, Joseph Grenny, Ron Mcmillan, and Al Switzler. 2007. *Influencer: The Power to Change Anything.* New York: McGraw-Hill.

Senge, Peter. 2006. *The Fifth Discipline: The Art and Practice of the Learning Organization,* 2nd revised ed. New York: Random House.

Shulz, Kathryn. 2011. *Being Wrong: Adventures in the Margin of Error: The Meaning of Error in an Age of Certainty.* London: Portobello Books Ltd.

❖ CHAPTER 12 ❖

Theory of Constraints (TOC)

The *Theory of Constraints* (*TOC*) is the conceptual framework behind *The Goal*, Eliyahu ("Eli") Goldratt's major bestseller. It has these main components:

- ◆ The Process of Ongoing Improvement (POOGI)
- ◆ Drum-Buffer-Rope (DBR)
- ◆ The Logical Thinking Process
- ◆ Critical Chain Project Management (CCPM)
- ◆ Throughput Accounting

These components can be used independently, and each part has its own band of experts, but as we'll see, there is a satisfying coherence to the whole, too.

Given the accessibility and popularity of Goldratt's writing, why are his methods not more widely applied? That question gives some context to the final part of this chapter, a look at the historical and ongoing relationships between TOC and Kanban.

TOC represents a large and important body of knowledge, not easily summarized. Bear that in mind as you read this overview. TOC material can be rather folksy, often featuring "Herbie" (a character from *The Goal*, constraint personified), "Murphy" (of Murphy's Law, used as a nickname for *variation*), and the like. At the other end of the scale is a rather intimidating soup of acronyms and technical terms unique to TOC. Sometimes both styles are used in the same source. Neither extreme is my style, and

I try here to explain TOC's key concepts and its most important terms in what I hope is clear language.

The Five Focusing Steps and the Process of Ongoing Improvement (POOGI)

TOC's *improvement cycle*, known as its *process of ongoing improvement* (*POOGI*), is laid out in the *five focusing steps*. It comprises four steps in a loop (the fifth step takes us back to the beginning):

1. **Identify** the system's constraint (the factor most responsible for the system's failure to achieve higher performance).

2. Decide how to **exploit** the system's constraint (make sure that the constrained part of the system is always working to capacity, never starved of anything that it needs).

3. **Subordinate** everything else to the previous decisions (realign the rest of the system to the needs of the constraint).

4. **Elevate** the system's constraint (find ways to increase capacity at the constraint).

5. **Repeat.** If in the previous steps the constraint has been broken, go back to step 1. **Warning:** Do not allow inertia to cause a system's constraint!

In this wording there are a few hidden assumptions that bear being made explicit:

♦ There is only one *constraint* that matters; working on other constraints would be wasteful. (Some wordings allow for a small number of significant constraints, but the language and this message tend to lose some impact.)

♦ The constraint typically takes the form of a *bottleneck* in the process—the step that limits the pace of the others in the process. You might see this referred to as the *capacity-constrained resource* (*CCR*). Occasionally, you might encounter a *non-instant availability* (*NIA*) resource, in which there is sufficient capacity overall but not necessarily at the time you need it (scheduled services are good examples of this).

♦ For the most part, the (economic) goal of the system is served most directly by improving *throughput*; considerations such as *inventory*, *lead time*, or *variation* tend to be secondary to that (though still important).

♦ Eventually, after removing internal constraints, the constraint will reside externally. When this happens, the system will typically be constrained by a supply-side bottleneck or by a shortfall of market demand relative to capacity.

These assumptions may be made open to challenge by a step 0:

0. Define the system's goal or objective.

For the purposes of this book, *POOGI-0* means POOGI with this step 0 explicitly included.

I have noticed that software projects are often managed under the assumption that the constraint is (and always will be) the development team. Specifically, keeping the developers busy is what counts, even when it is clear that high-quality requirements are in short supply, or that the team's delivery capability is not being matched by the customer's ability to realize the planned benefits of the new functionality they receive. Only by challenging these assumptions can we be sure that management attention is in the right place.

Drum-Buffer-Rope (DBR)

Drum-Buffer-Rope is the production scheduling system of the Theory of Constraints. As such, DBR plays the role in TOC that kanban systems do in the Kanban Method.

Assume that with step 1 of the five focusing steps we have *identified* a constraint activity at some late stage in some overall process, perhaps an integration or inspection point. If no such constraint can be found, TOC practitioners might treat the final shipping activity as the constraint and apply DBR there.

The **drum** (think "drum beat") is a work schedule for the constraint, planned ahead of time. Step 2 of the five focusing steps tells us to *exploit* the constraint, so we plan to keep the constraint busy, at a high but sustainable *utilization*.

The **buffer** is the time allowed for each kind of work item, component, or subassembly to reach the constraint via the upstream process. During this period, we apply *buffer management*, meaning:

♦ Monitoring these work products through the process upstream of the constraint;

♦ Assigning a red, amber, or green (RAG) status to each item according to how much of its time buffer has been consumed;

♦ Marshaling toward the constraint the most time-critical pieces, that is, those with a red RAG status. Practices such as *daily standup meetings* and *swarming* that we may think of as belonging to Kanban or Scrum belong in TOC, too.

Buffer management effectively *subordinates* the upstream process to the needs of the constraint (step 3 of the five focusing steps). Successful buffer management ensures that there is always a supply of work ready for processing at the constraint.[41]

The **rope** ties the buffer's input to its output. Work is released into the system according to a schedule constructed relative to the drum—the work schedule for the constraint activity. The rope makes DBR an interesting kind of *pull system*; work is pulled into the system in good time for it to be processed at the constraint at the planned time.

In effect, DBR has two schedules, one downstream at the constraint and another at some upstream point. Rather than plan in detail what happens between those two points, work is managed through the process day to day, prioritizing work items according to their time-criticality. Given a sustainable pace through the constraint and a sensible upstream buffer, a stable and predictable flow should result.

The Thinking Processes

Like Kanban, TOC recognizes that improvement implies change, and that people find change difficult. TOC includes a suite of thinking tools designed to address what it calls *resistance to change*.

---

41. Note that Kanban and TOC mean slightly different things by "buffer"; you might offend a TOC purist if you claim to implement DBR simply by maintaining a buffer of WIP ahead of a bottleneck. In Kanban, a buffer is a queue whose purpose is to provide a steady supply of work. In TOC, the buffer is a defined period of time.

TOC describes a number of *layers of resistance*. The number and exact wording of these layers has evolved over time; Kelvyn Youngman has traced through this evolution on his excellent website, *A Guide to Implementing the Theory of Constraints (TOC)* (www.dbrmfg.co.nz). The current nine-layer model shown in Table 12.1 is credited to Efrat Goldratt.

Table 12.1   Efrat Goldratt's model of the nine layers of resistance to change

Resistance	Tool
There is no problem	
Don't agree about the extent of the problem	Current Reality Tree
Don't agree about the nature of the problem	Cloud
Don't agree about the direction of the solution	Injected Cloud
Don't agree about the completeness of the solution	Future Reality Tree
Can see additional negative outcomes	Negative Branch Reservation
Can see obstacles	Pre-requisite Tree
It's not exactly clear how to implement the solution	Transition Tree

Faced with a specific resistance in the left-hand column of Table 12.1, TOC practitioners will turn to the corresponding tool in the right-hand column. With these, the practitioner describes and manipulates the problem space graphically, searching for ways to resolve the issue. These tools plus a couple of others complete TOC's *Thinking Processes.*

The goal of the Thinking Processes is to define a target state that can be delivered through a series of transformations. TOC covers the delivery part of this part, too, with its own project-management framework, *Critical Chain Project Management (CCPM).*

Critical Chain Project Management (CCPM)

Critical Chain Project Management (CCPM) applies the thinking of Drum-Buffer-Rope to the problem of planning and controlling projects. In outline:

1. Start with the network of project tasks, their estimates, and dependencies.

2. Schedule tasks so that:

- ◆ No task is started before its dependencies are fully satisfied (no "non-standard" dependencies between tasks).
- ◆ Multitasking is eliminated (no "leveling" of "resources"[42] across tasks).
- ◆ Project duration is minimized (to a first approximation).

3. Identify the *critical chain*—the sequence of tasks that determines the project's overall duration.

4. Separate task estimates into two components, the expected duration and the remaining safety margin. Typically this is done by a simple rule of thumb, such as assuming that the safety margin in each estimate makes up one-third of the total.

Move all of the safety margins to a *buffer*. These live either at the end of the project (the *project completion buffer*) or protecting each dependency (*feeding buffers*).

Figure 12.1 shows a simple plan with its project completion buffer and one feeding buffer.

Step 2 of the CCPM is analogous to the *drum* in DBR; it should generate a rough plan that is aggressive without making unsafe assumptions. DBR explicitly avoids some of the "plan compression" tricks that get project managers into trouble.

Corresponding to the *rope* of DBR, work is scheduled to start just as its corresponding buffer would start to be consumed.

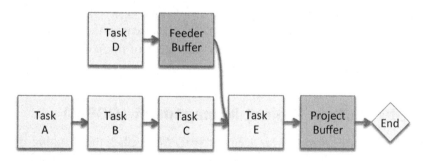

Figure 12.1   A simple plan with its project completion buffer and one feeding buffer

---

42. No, I don't like "resources" as PM-speak for "people" either, but we use the word in a general sense here.

Once work is underway, *buffer management* takes over, just as in DBR. *Buffer penetration charts* might aid the project manager in this; these show buffer consumption charted against time or against project progress. An interesting variant of these are *fever charts* (Figure 12.2); their colored backgrounds allow each buffer's RAG status to be read directly off the chart.

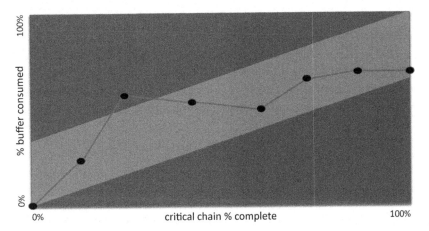

Figure 12.2   A fever chart

Throughput Accounting

TOC has its own accounting model, *Throughput Accounting*, a detailed treatment of which is beyond the scope of this book. It aims to reverse these supposed priorities of traditional management that is driven by the conventional *Cost Accounting* model:

1. Reduce cost.
2. Reduce required investment or inventory.
3. Increase throughput, defined here as sales less *totally variable cost* (*TVC*) per unit of time.

The logic of Throughput Accounting is that reductions in cost, capital investment, or inventory that look good on paper in the Cost Accounting model may be damaging to throughput, and therefore detrimental to the interests of the organization.

TOC's Relationship to Kanban

Chapter 8 refers briefly to a story from the "blue book"—David Anderson's Kanban book—in which an interesting transition unfolds. His chapter, "From Worst to Best in Five Quarters," describes what was to become Kanban's first case study. It tells the story of how Microsoft's XIT Sustained Engineering Team was transformed out of all recognition, achieved through the implementation of a pull system and by a series of incremental improvements.

This pull system was first conceived as a simple Drum-Buffer-Rope system (David's first book was heavily TOC-based); only later was it reinterpreted as a kanban system. And along the way, the Kanban Method was born. We seem to give much more credit now to Lean (Chapter 13) than we do to TOC; is that fair?

To answer that, let's review the main elements of TOC.

POOGI and the Five Focusing Steps

Like many Kanban trainers, I like to reference POOGI and the Five Focusing Steps when I teach Kanban. I particularly remember one class in which a small company's entire middle management layer was in attendance. It dawned on us that if the company had a constraint (and surely it must), it had to be represented by one of those managers in the room.

All eyes turned to the finance manager. Everyone saw how important it was that she got what she needed when she needed it. Her colleagues resolved to remove from her and her small team its overhead of chasing for paperwork, clarifications, and corrections. Special priority would be given to activities that brought in cash. Everyone in the room determined that things were going to be better, not just for that team but for the company as a whole.

That's a great story, but Kanban is not "POOGI with kanban boards." The constraint—typically presented as some kind of bottleneck—is rarely the Kanban practitioner's first line of attack. Get past the delicate issue that no person or team wants to be labeled a bottleneck, and you still have the problem of identifying them. How sure can you be of the location of your bottlenecks when WIP is high, there are orders-of-magnitude differences in lead times between work items, people can easily move between activities, and (give or take) everyone gives the impression of being equally busy?

Perhaps *bottleneck* should be added to that list of tools, concepts, and metaphors that don't translate from manufacturing work to knowledge work quite as readily as some would have you believe. Chapter 14 identifies some more of those.

The bottleneck may not enjoy a first-class status in the Kanban Method, but POOGI will long continue to be taught. It promotes **understanding**, and there is still power in the idea that you need to keep on identifying and addressing your system's constraints, especially when your mind is open to the possibility of much broader constraints—lack of knowledge, feedback, learning, trust, and so on, or the attachment to unhelpful ways of thinking.

DBR and CCPM

Kanban systems have of course replaced DBR as the scheduling tool of choice in the Kanban Method. Both are pull systems, both can be effective. What separates them isn't so much the visualization; rather, it is the manner of introducing them. When you implement DBR, you are fundamentally changing the way you schedule and control the flow of work. Although it would be wrong to suggest that kanban systems aren't at all disruptive (after all, we want them to provoke change), they're typically laid on top of existing processes.

Some within the Kanban community see both DBR and CCPM as complementary to Kanban (see, for example, my friend Steve Tendon's book in the list of further reading), and certainly there are some interesting parallels that are well worth exploring. The rope concept suggests *explicit policies* for governing how fixed-date work is started. Buffer management adds a risk-management flavor to *manage flow*, perhaps to be augmented by fever charts and related metrics. If you are faced with larger, traditionally managed projects, take a look at how CCPM accommodates variation, especially with regard to dependencies.

Not that I believe that knowledge work is usually best done in project-sized chunks or that work items should be treated as date-driven by default; let's reserve the specialized tools for that minority of cases that really deserve them.

The Thinking Processes

"*Layers of resistance*" and "*overcoming resistance to change*"? Those phrases might themselves generate resistance, but that's easily fixed (Youngman suggests "*degrees of acceptance*"), and the categorizations do seem helpful.

The Thinking Processes as thinking tools? I can't claim to be a skilled practitioner myself, but I've seen some impressive outputs from others. One day I might be tempted to learn how to use them.

But again, some caveats. At issue is not the potential power of the tools but their intent. Kanban is about growing a learning capability inside your organization. You don't achieve that by following the transition plan of an outside expert or an internal specialist. A target state will keep you competitive only for a while; in the longer run, learning wins. To my mind, therefore, project-based approaches to change aren't the sustainable answer.

But Still . . .

Don't be too put off by my cautions and caveats—TOC might not be a foundation of today's Kanban Method, but its role as a source of insight and inspiration is secure. Do read *The Goal*—it won't take long and you will be rewarded. Look further into the Theory of Constraints and expect to be challenged.

Further Reading

Goldratt, Eliyahu M. and Jeff Cox. 2004. *The Goal: A Process of Ongoing Improvement*, 3rd ed. Great Barrington, MA: North River Press.

Tendon, Steve and Wolfram Müller. 2013. *Tame the Flow*. leanpub.com/tame-the-flow.

Youngman, Kelvyn. *A Guide to Implementing the Theory of Constraints (TOC)*. www.dbrmfg.co.nz.

❖ CHAPTER 13 ❖

Agile

We are uncovering better ways of developing software by doing it and helping others do it. Through this work we have come to value:

Individuals and interactions *over* **Processes and tools**

Working software *over* **Comprehensive documentation**

Customer collaboration *over* **Contract negotiation**

Responding to change *over* **Following a plan**

That is, while there is value in the items on the right, we value the items on the left more.

Kent Beck, James Grenning, Robert C. Martin, Mike Beedle, Jim Highsmith, Steve Mellor, Arie van Bennekom, Andrew Hunt, Ken Schwaber, Alistair Cockburn, Ron Jeffries, Jeff Sutherland, Ward Cunningham, Jon Kern, Dave Thomas, Martin Fowler, Brian Marick

© 2001, the above authors. This declaration may be freely copied in any form, but only in its entirety through this notice.

So reads the Agile Manifesto, the product of a landmark meeting in February 2001 at the Snowbird ski resort in Utah. A movement was born.

I will come clean: When I identified **collaboration** as one of Kanban's values I was making a conscious nod in the direction of the Agile Manifesto. It's easy to read other values into the manifesto, too: **Respect, flow,** and **customer focus** spring most immediately to mind. There are some more subtle resonances in "Responding to change over Following a plan," which hints at adaptability through *evolutionary delivery.* Evolution of method gets an indirect mention, not as a value but in the preamble, in "uncovering better ways of developing software."

Further evidence of these and other values can found in this addendum to the manifesto, the *Twelve Agile Principles*:

The Agile Manifesto is based on twelve principles:

1. *Customer satisfaction by rapid delivery of useful software*
2. *Welcome changing requirements, even late in development*
3. *Working software is delivered frequently (weeks rather than months)*
4. *Working software is the principal measure of progress*
5. *Sustainable development, able to maintain a constant pace*
6. *Close, daily cooperation between business people and developers*
7. *Face-to-face conversation is the best form of communication (co-location)*
8. *Projects are built around motivated individuals, who should be trusted*
9. *Continuous attention to technical excellence and good design*
10. *Simplicity—the art of maximizing the amount of work not done—is essential*
11. *Self-organizing teams*
12. *Regular adaptation to changing circumstances*

Rather than analyze these in the abstract, let's look at some of those "better ways," three methods that to various degrees provided some of the context to the Snowbird meeting. They're interesting models in their own right.

Three Agile Methods

Feature-Driven Development (FDD)

Jeff de Luca first developed FDD in 1997 for a specific project at a bank in Singapore. As documented in the "blue book," it enters the Kanban story in 2004 when David Anderson introduced it at Motorola. In overview, FDD's process is diagrammed in Figure 13.1.

As with all Agile methods, FDD should not be mistaken for a traditional, phased project process. The outputs from the build activity emerge in feature-sized increments over the course of the project. Working backward, each build is preceded by design, which happens feature-by-feature.

Planning (*Plan by feature* preceded by *Build a features list*) is performed as needed, in chunks of no more than a few weeks' worth of work.

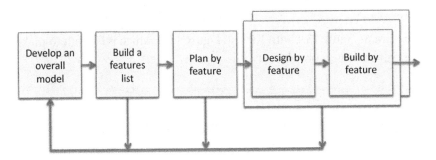

Figure 13.1   Feature-driven development

One of the most notable features of FDD is its first activity, *Develop an overall model*, comprising a technical model (an *object model*, likely expressed in the graphical notation of the *Unified Modeling Language, UML*) and accompanying notes. Initially this model is very high level; it evolves and is refined "*just enough for now*" in light of feedback from the other four activities, the *Design by Feature* activity most especially.

What is not clear from the diagram is that the project's *clients* (FDD's term for customers and other project stakeholders) participate in the modeling activity right through the life of the project. The key Agile elements are there: It is collaborative, adaptive, and incremental. It acknowledges that knowledge must be accumulated and re-evaluated over time and tested by interaction with the real world.

Extreme Programming (XP)

In late 1999 or early 2000, a newly published book with a strange title arrived unexpectedly on my desk via the internal mail, courtesy of my then-manager, Thomas (Thommi) Suessli. It was Kent Beck's *Extreme Programming Explained: Embrace Change*. I have to admit that I was a little bemused by it to start with, but I stuck with it, later investing in other books in the series.

Like FDD, XP predates the Agile Manifesto. It recognizes five values: **communication, simplicity, feedback, courage**, and **respect** (a later addition).

I find no direct analog for simplicity in Kanban's nine values (not that there is any conflict), but communication and feedback correspond well with **collaboration** and **transparency**; there is certainly a relationship between courage and **leadership**; and **respect,** of course, maps directly.

The differences between XP and FDD are striking. Compare FDD's diagram in Figure 13.1 with XP's, in Figure 13.2, which is laid out to make the comparison more obvious.

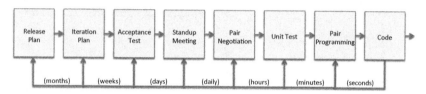

Figure 13.2   The XP process

Several differences stand out:

◆ There is no modeling activity. Moreover, the tightest feedback loops are to the right of the picture (around *Code*), not the left.

◆ Neither is there an explicit design activity (but there are a number of other activities between planning and programming). In XP, design is *emergent.*

◆ Two activities are test-related: *Acceptance Test* (curiously, this appears much earlier in the process than you might expect) and *Unit Test* (where code-related tests are written just ahead of the code they help to exercise).

◆ Two activities involve pairs: *Pair Negotiation* and *Pair Programming.*

The clue is in the name: XP's heart really is in the programming. Why start with model-building when the code itself can be the model? The programming is made "extreme" by the removal of anything extraneous, the intense collaboration of pairing, and the tightest of feedback loops.

My favorite XP-inspired sound bite goes like this:

If it hurts, do it more frequently, and bring the pain forward.[43]

---

43. There are several versions of this line. My source for this particular version is the excellent book by Jez Humble and David Farley, *Continuous Delivery* (2010. Upper Saddle River, NJ: Addison-Wesley).

This sounds crazy, but it's inspired! Not sure how to test your code? Write the tests first. Finding acceptance testing painful? Integrate it into the product design process. Does deployment hurt? Plan to deploy very much more frequently (continuously, even). XP is built on the key insight that these sources of pain are actually opportunities for highly valuable feedback; it then has the courage to turn the dials up to maximum.

This pursuit of fast feedback has fuelled rapid development in low-level practices and tools. New communities have spawned around them, refining them further and working hard to share them more widely. Rates of both adoption and innovation have been high, supported by the *open-source* nature of their development. Some of this would have happened anyway, but there's no doubt that XP played a significant role in this.

Before he created XP, Kent Beck was already a unit-testing pioneer, founding the *xUnit* family of *open-source* automated unit testing frameworks with SUnit, an implementation for the Smalltalk language. Since that time, xUnit implementations for many other languages have been created—JUnit for Java and Test::Unit for Ruby, for example. Unit testing isn't just a solved problem now; it's mainstream, with *test-driven development* (*TDD*) heading that way, too.

Automated acceptance testing involves the following:

1. The specification of expected product behavior—often in forms capable of being read (and even written) by non-developers

2. The technical means to interact with the product without human intervention, often via (or by simulating) a web browser

3. Some glue between the two—programming language support and so on

There has been plenty of innovation here, too, spawning whole communities around the various frameworks and specification techniques. Some, such as *behavior-driven development* (*BDD*), have grown to become methodologies in their own right.

Open-source *distributed version control systems* (*DVCS*) such as *git* have allowed large projects such as Linux to be worked on by hundreds (if not thousands) of developers—truly a huge collaborative effort. These tools have gone on to form the foundations of several solutions to the deployment problem also. Increasingly—and for a part-time coder like me, wonderfully—these are built into the hosting platform; getting my latest creation online can now be as simple as a quick "git push" from the command line. *Continuous delivery* takes the automation of code management, testing, and deployment to the extreme; firms such as Amazon now make such frequent releases that mean time between them is measured in seconds!

With entire workflows taking just hours or days, work items in XP need to be both small and testable. Finding *functional requirements, use cases,* and *features* too large, XP popularized the *user story*, a requirement expressed on a card as a single-sentence conversation starter—often, stereotypically ("As a <user> I want <something> so that <benefit>"). These, too, have been the subject of much exploration—whole books have been written on the subject.

Scrum

You may have noticed that I neglected to expand on the planning aspects of XP. There's a reason for that: XP's *planning game* has, to a very significant extent, been supplanted by Scrum. To many people, Agile is now almost synonymous with Scrum, with or without the technical practices of XP.[44]

Scrum is an iterative and incremental process framework developed in the mid-1990s (like XP, it predates the Agile manifesto). Its process is depicted in Figure 13.3.

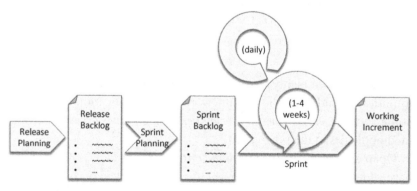

Figure 13.3   The Scrum process

---

44. With some justification, Martin Fowler famously coined the label "Flaccid Scrum" for Scrum devoid of XP's technical practices. See http://martinfowler.com/bliki/FlaccidScrum.html.

Conveniently, Scrum can (just about) be described in three threes:[45]

1. Three roles, the *Product Owner* (PO), *Scrum Master* (SM), and *Team*

2. Three events, *Sprint Planning* before the sprint, the *Daily Scrum* each day during the sprint, and two meetings to close the sprint, namely the *Sprint Review* and the *Sprint Retrospective*

3. Three key artifacts, the *Product Backlog*, the *Sprint Backlog*, and the *Burndown Chart*

The Scrum Master's role is to facilitate development and improvement on behalf of the team and the Product Owner. The Product Owner represents the product being developed, often acting as a proxy for its customers. At the center of it all—and very much by design—is the team. Note that the SM and PO tread a fine line between protection (giving the team the space to be productive) and insulation (where collaboration is only either internal or intermediated).

The three events are meant to be held regularly, like clockwork. The Sprint Planning meeting decides the content of the next increment, that is, the work to be delivered in the next sprint, which is typically a period of two weeks. The Daily Scrum is a standup meeting (Chapter 1 touches on this). The Sprint Review looks at the work that was completed (demonstrating it to stakeholders) and the planned work that was not completed. The Sprint Retrospective looks back on the completed sprint to determine what went well and what might be improved.

Scrum's rhythm is both a key strength and (deliberately) its biggest source of challenge. The hard part is selecting and then delivering the right quantity of work so that it fits snugly into the *timebox* of the sprint. When there is too much or too little, the result is the same: waste and frustration both inside and outside the team.

The Product Backlog and Sprint Backlog describe at various levels the work still to be done; Burndown Charts (a cousin of the cumulative flow diagram) visualize progress in the sprint or toward the next release. Typically, these artifacts are kept highly visible.

Many of these elements could be described as pragmatic solutions to common problems. Do you have multiple customers, far removed from

---

45. That the Kanban values also come in three groups of three is a coincidence!

the team? Have a highly available Product Owner. Do you need someone to model and facilitate appropriate practices and behaviors? Bring in a Scrum Master. Are your feedback loops too slow relative to the rate of environmental change? Plan your work to fill short timeboxes and meet daily.

Here, the main elements of Scrum are described separately, but its designers expect its adopters to make them work all together. This isn't easy. It isn't meant to be easy; to quote Ken Schwaber, "*Scrum is hard and disruptive.*"[46] Significant change may be demanded from the organization in order to make it work effectively.

Kanban and Agile

Compatibility

We're often asked, "Is Kanban Agile?" To anyone who understands Kanban as the *start with what you do now* method, that's a slightly odd question, but still, it deserves a respectful response.

When "Agile" is used as an adjective like that, it's worth drilling down a bit to find out what's really meant by the question:

♦ Are the values of Kanban and Agile compatible? Yes, absolutely! There is a basis here not just for comparison but for integration.

♦ Can Kanban help "improve agility" or make things "more agile," improving an existing software development process—explicitly Agile or otherwise—in directions entirely consistent with Agile values and principles? Not only is the answer to that question a resounding "yes," it is what the method was first developed to do.

♦ Is Kanban, in the words of the manifesto, a *way of developing software?* No. We are splitting hairs, perhaps, but in this sense it is misleading to describe Kanban as an Agile method. It isn't a development process (or any other kind of delivery process) at all; there is nothing in the definition of the method that ties it to software. It is a management method that is broadly applicable to creative knowledge work, with a particular focus on organizational change.

♦ Is Kanban part of the Agile movement? Kanban's community identity makes sense both inside and outside the Agile movement (note

46. http://www.controlchaos.com/storage/scrum-articles/Scrum Is Hard and Disruptive.pdf

that both communities have a similar relationship with Lean). Some people seem to be troubled by that ambiguity, but actually it's helpful and necessary. Ideas find room to grow and flourish in their own communities, and when the time is right, there is ample opportunity for cross-fertilization. And to providers of supporting services (trainers, coaches, consultants, and so on), the ability to choose from multiple identities can be very convenient.

At the level of methods and practices:

◆ Does Kanban work with iterative methods, Scrum in particular? Yes, and in the case of Scrum, the combination even has a name—*Scrumban* (described in Chapter 16). Kanban can work both inside Scrum, where it mainly drives team-level improvement, and outside it, where it helps a deliberately team-centric framework address the challenges of scale.

◆ Doesn't Kanban mean abandoning iterations and other elements of Scrum? This is a serious misconception. Kanban is the *start with what you do now* method; we would be the first to warn you not to drop aspects of your current process in an uncontrolled fashion. However, it would be dishonest of us to pretend that your pursuit of flow won't at some point test your commitment to timeboxes, story points, and the like. How you and your organization deal with that will be a matter of choice.

◆ Do Agile and its methods and practices represent important models for Kanban practitioners? Most definitely yes—it seems almost unthinkable that an effective Kanban practitioner working in the software development domain would get very far without a deep understanding and respect for Agile.

When to Use Kanban

One question remains: In an Agile context, when and how might Kanban help? More questions about your current situation will help answer that:

◆ Is Agile's principle of *sustainable pace* still just an aspiration? Are people still overburdened in a process that doesn't seem to fit all that well? Could Kanban-style **transparency** (Chapter 1) and **balance** (Chapter 2) provide some relief?

- ◆ Is your **collaboration** (Chapter 3) focused mainly inwardly? Does **customer focus** (Chapter 4) suffer as a result of over-protective intermediation around the team or of excessive internal focus on the technology, the product, or the team?
- ◆ Are team-centric and process-centric approaches failing to deliver needed gains in end-to-end **flow** (Chapter 5)? Are local gains even making things worse elsewhere?
- ◆ Are **leadership** (Chapter 6), **understanding** (Chapter 7), and **agreement** (Chapter 8) under-appreciated? Is **respect** (Chapter 9) too easily forgotten when the big decisions are being made?

It is hard to change how organizations tick. Agile adoptions face that challenge all the time, and it should come as no surprise that issues such as these arise. Whether you are just planning to set out down that road or are already well along the way, the *start with what you do* method is there to help, not to judge.

If your wider organization is ready to make the kind of changes called for by a *hard and disruptive* Agile adoption and you can be very sure of success, you might not need Kanban. For organizations unprepared to take that risk, Kanban offers an alternative path to agility, an open-ended journey of co-evolution in which better ways of doing things are waiting to be discovered.

The Agile Model

Don't be fooled by this necessarily process-centric treatment of FDD, XP, and Scrum. With any of these methods you can go through all the iterative motions and still find that:

- ◆ People serve the process, not the other way around.
- ◆ The product is driven by the loudest internal voices, not the emerging needs of actual customers.
- ◆ Lots of work gets done without the end product ever being used for real.
- ◆ Changes in direction cannot be contemplated, let alone accommodated.

On its own, process gets you only so far. If the aim is to be Agile, the Agile values need to be evident. Technicalities aside, if it's working neither for the team nor for the customer, call it something else!

Agile has been and still is a game-changer. It has legitimized evolutionary delivery, wresting control of swathes of the software industry away from a plan-driven style of project management that was often ill-equipped to deal with uncertainty. No self-respecting Kanban practitioner can afford to ignore it.

Further Reading

Manifesto for Agile Software Development (agilemanifesto.org)

Ambler, Scott. *Feature Driven Design (FDD) and Agile Modeling.* www.agilemodeling.com/essays/fdd.htm.

Beck, Kent. 1999. *Extreme Programming Explained: Embrace Change.* Upper Saddle River, NJ: Addison-Wesley.

Humble, Jez and David Farley. 2010. *Continuous Delivery: Reliable Software Releases through Build, Test, and Deployment Automation.* Upper Saddle River, NJ: Addison-Wesley Signature Series (Fowler).

Rubin, Kenneth S. 2012. *Essential Scrum: A Practical Guide to the Most Popular Agile Process.* Upper Saddle River, NJ:Addison-Wesley.

Schwaber, Ken and Jeff Sutherland. *The Scrum Guide.* www.scrum.org/Scrum-Guide.

Sheridan, Richard. 2013. *Joy, Inc.: How We Built a Workplace People Love.* New York: Portfolio Penguin.

❖ CHAPTER 14 ❖

TPS and Lean

I n 1978, the year of his retirement from Toyota, executive vice president Taiichi Ohno published a book describing the *Toyota Production System* (TPS). His book shared a remarkable number of concepts and tools that were barely known outside of Japan, such as:

- ◆ *Just-in-time (JIT)*—the radical idea that the right materials, parts, and assemblies should arrive where they're needed only as they're needed, and in the smallest possible quantities
- ◆ *Autonomation*—"automation with a human touch," the production line's early warning system
- ◆ The *andon* system—a visual indication of trouble combined with the means for ordinary shop-floor workers to "stop the line"
- ◆ The *Five Whys (5W)*—a technique for root cause analysis
- ◆ *Kanban*—the card-based system by which just-in-time production is managed

Over the following decade, interest in Toyota grew as its position in the global car market strengthened. In 1998 an English translation of this seminal book at last became available. Two things strike you as you read it: the brilliant simplicity of the system and its tools, and Ohno's remarkable strength of purpose as he strove to develop a system capable of realizing the vision of the company's founders, the Toyoda family.

Not that most of us were reading Ohno back then. The book that generated popular interest in Toyota's unusual methods and brought the term *Lean Production* into the mainstream was *The Machine That Changed the World* by James P. Womack, Daniel T. Jones, and Daniel Roos, published in 1991. Two years later Womack and Jones followed this up with *Lean Thinking*. Japan had Toyota; the world had Lean.

Another raft of Japanese words entered the lexicon:

+ *Kaizen*—continuous improvement through incremental change
+ *Kaikaku*—radical change
+ *Heijunka*—production leveling; that is, deliberately mixing work on the production line rather than producing similar work in batches
+ *Poka-yoke* (or *Baka Yoke*)—mistake-proofing
+ *Gemba, gembatsu, genjitsu*—the "three reals," the real place where work is done, the real thing, and real facts, respectively
+ *Hoshin Kanri*—strategic planning and policy deployment

And the list goes on. In retrospect, the early years of Lean did a great job of packaging up Toyota's tools and techniques for western consumption but somehow failed to give much insight into the thinking that generated them. This began to change as writers such as John Shook, Mike Rother, and Steven J. Spear focused on Toyota's approach to management.

Three Lean Tools

A detailed explanation of how a production line works at Toyota is beyond the scope of this book. With just enough detail to allow comparisons to be made with the kinds of kanban systems described in Part I, here is a simple setup that uses *kanban, heijunka*, and *andon*:

+ No parts or materials get supplied, and nothing—neither subassemblies nor the finished product—gets built without an appropriate *kanban*.
+ At the delivery end of the production line, *kanban* are[47] pulled one at a time from the *heijunka* box. Slots in this box (or rack) organize the *kanban* in two dimensions, with time of day along one axis and product type along the other. By choice and by physical constraint, production of each product type must be spread across the day.
+ The workstation (perhaps a *cell* with multiple workers) at the front of the production line assembles and packs the order to the specifi-

---

47. "Kanban are like sheep." One kanban, two kanban, . . .

cation given on the *kanban*. As the supply of parts here drops below a replenishment level, *kanban* are sent upstream.

♦ And so the process continues, subassemblies moving down the production line, *kanban* moving upstream all the way up to Goods Inwards, where parts and materials must be ordered from outside suppliers . . .

♦ . . . until someone pulls the *andon* cord because they've spotted a problem; alarm lights come on and the line comes to a halt. After investigating and rectifying the problem, production starts up again.

This description is rather simplistic, but there's enough here for some striking characteristics of the system to be noted:

All three tools (*kanban, heijunka,* and *andon*) are examples of *visual management*.

♦ Inventory of all kinds is limited. Neither basic supplies nor WIP (the subassemblies, or partially-built products) will be replenished until equivalent amounts have been pulled from downstream.

♦ Even though, perhaps, it might seem more efficient to do so, the production line doesn't work in large batches of similar items. Instead, it produces a variety of products spread over the course of the day.

♦ Workers on the production line would rather *stop the line* (for everyone) than allow work of inferior quality to proceed.

♦ This system and the kanban boards from Part I work very differently. On the production line, the *kanban* are sent upstream to signal that there is demand to be fulfilled. On our boards, the cards represent work items as they flow downstream; signals are implied by the gaps between the actual amount of work in progress in each state and the corresponding WIP limits.

♦ The *heijunka* box and our kanban boards both allow the mix of work to be managed.

It seems perverse, not only setting things up to work in this deliberately difficult and seemingly inefficient manner, but empowering workers to bring it all to a halt at any time! Clearly there must something special about the company's culture for this to work at all, but why would they choose to do things this way?

TPS and Lean in Perspective

To answer that question you must understand TPS as a magnificent example of Systems Thinking.

It starts with a vision, a *true north* that gives the direction for change:

◆ *Single-piece flow, in sequence, on demand, with zero defects; 100% value-adding activities and security for the people performing them*

The technology does not yet exist to make it economical to run the entire production line in batches of one (which is what single-piece flow means), but the pursuit of this perhaps impossible vision is what propelled Toyota from its struggles in postwar Japan—where land, factory space, plant, and materials were all in short supply—to the global market leadership position that it now occupies.

The tools support one or both of two purposes:

1. Satisfying customer demand as quickly and as smoothly as (currently) possible with the minimum amount of inventory

2. Evolving the company to take it closer to its vision, harnessing the abilities of its entire workforce to smooth flow, reduce inventories, prevent defects, eliminate other forms of waste, and (not least) design new products that customers really want and that can be produced both profitably and sustainably

The two pillars of *just-in-time* and *respect for people* are shorthand for those sub-goals.

Often missed is this crucial point: The pillars and the tools can be seen in their proper perspective only once it is grasped that Toyota's pursuit of perfection is a multi-generational challenge. Toyota works not only to build cars, but also to build the company capable of delivering on its vision.

Divorced from that kind of thinking, the tools of Lean can seem shallow. Without the tools, it can be even worse—too often we hear Lean reduced simply to a short-term focus on waste (perhaps to dress up exercises in cost cutting), or to continuous improvement (important, but very hard to sustain in isolation). The challenge of the Lean movement is to make sure that the thinking is packaged up with the tools so that people can apply them appropriately in context.

Lean Improvement

Much of that Lean thinking is built into and around these five improvement steps:

1. **Identify value** from the customer's standpoint.
2. **Identify the value stream**—the value-creating steps in the process —and seek to eliminate what is non-value-adding.
3. **Create flow**, removing delays between those value-creating steps, seeking smoothness.
4. **Establish pull**, where work is taken from upstream only in response to downstream demand, ultimately from the customer.
5. **Identify waste**, removing impediments to smooth flow, reducing delays, reducing inventories, eliminating defects at the source, and so on.

These steps are often referred to as *Lean principles* (a slight misnomer perhaps, but never mind). Applied repeatedly, we have an improvement cycle analogous to the POOGI loop of Theory of Constraints (Chapter 12), replacing POOGI's specific focus on constraints (usually understood to be constraints on throughput) with attention to delays, smoothness, and WIP.

Inside that improvement cycle, Lean (and Toyota before it) embraces PDCA (Chapter 3) as the way to frame each incremental change. Larger changes may be documented and planned in an *A3* (named after the paper size), and developed over a period of time in the context of a mentor/mentee relationship of some kind. Smaller changes might be managed through various styles of structured dialog known as *katas* (another Japanese word, imported this time not from Toyota but from the martial arts; it means a choreographed pattern or routine).

Drilling down into steps 2 and 5 of the Lean principles, Ohno identified *seven wastes,* or non-value-adding activities:

1. **Transportation**—A source of delay, cost, and risk of loss or damage

2. **Inventory**—materials, work-in-progress (WIP), finished but undelivered goods—wasteful both for the delays incurred and for the cost of financing it

3. **Motion**—damage to people and equipment caused by the production process

4. **Waiting**—time spent by work items in inactive states

5. **Over-processing**—doing more work than is necessary to meet specifications

6. **Over-production**—producing work in excess of immediate demand

7. **Defects**—effective capacity wasted on bringing inferior work up to the required standard

Taken as an integrated whole, the Lean principles, the wastes, the visual management tools (*kanban, heijunka, andon,* etc.), and the management practices (*kaizen,* A3, *hoshin kanri,* etc.), are far from shallow. Not only are there few systems as well documented as this, it takes time and effort to fully appreciate the thinking that went into it and the theory that underpins it. In short, Lean is a significant body of knowledge.

Lean Product Development

The Lean thinking described here so far still leaves a significant gap. Bluntly, Toyota's true north will never be reached on a diet of improvement alone. "*On demand*" and "*security for people*" need more than just production excellence—they need eager customers, a pipeline of good products, and sustainable profits.

Lean Product Development (LPD) seeks to address this gap. Relative to Lean manufacturing, it is quite young, expanding at a rate that allows for significant diversity. Some of the different approaches to LPD are represented here by a selection of its authors. This is by no means an exhaustive survey, more a list of some of LPD's key thinkers who have influenced Kanban's development:

◆ Donald G. Reinertsen takes Lean back to its first principles in queuing theory and economics, allowing the ideas and techniques of both Lean manufacturing and Lean Product Development to be applied outside their original domains much more effectively. The influence of Don's book *The Principles of Product Development Flow: Second Generation Lean Product Development* extends well beyond the Lean community. The Kanban Method might not exist were it not for Don's personal interest in David Anderson's work; *classes of service* and *cost of delay* are two Kanban concepts directly attributable to him.

◆ Michael N. Kennedy is the author of *Product Development for the Lean Enterprise* and *Ready, Set, Dominate*. The "Set" in the latter title refers to *set-based learning*, highlighting a crucial difference between how Toyota manages manufacturing and product development (with TPS and TPD respectively). Instead of the one-thing-at-a-time approach of single-piece flow, set-based approaches organize the parallel search for solutions that capture the appropriate combination of technical feasibility, cost-effectiveness, and attractiveness. In software development, where physical limitations are usually unimportant, we instead like to *defer commitment*, not so much exploring the solution space as allowing for multiple possibilities, buying flexibility for little or no cost.

◆ Peter Middleton and James Sutton weren't the first writers to consider the applicability of Lean principles to software development; neither were they the first to call out the inappropriateness of mass production as a metaphor for creative knowledge work. However, their book *Lean Software Strategies* is commendable for its commitment to customer needs, asserting that a deep understanding of the most lasting customer needs (which they refer to as *values*) leads not just to more fruitful relationships but to more enduring products and product architectures as well.

◆ Last but by no means least, Mary and Tom Poppendieck took the Lean principles, the wastes, and other concepts from Lean manufacturing and mapped them to a set of principles more appropriate to software development. There's no doubt that their books[48] had a significant impact on both Agile and more traditional thinking long before LPD fully established itself as a discipline.

---

48. Adding to the previous chapter's personal note on influential books and former line managers, Dr. Peter Lappo, who holds the unique distinction of having been both my line manager and my wife's (not simultaneously), lent me a copy of the Poppendiecks' first book soon after its publication in 2003.

It seems safe to assume that LPD will continue to develop along pragmatic lines, integrating the following:

1. Practices extracted from Toyota and other leading companies, made sufficiently separable from their host companies that they can be adopted elsewhere with some confidence

2. Practices developed and found to work well in particular contexts, some of which may with time turn out to be more broadly applicable, some not

3. A strong theoretical underpinning, some of it extracted from the previously mentioned practices, some of it imported from other fields, the rest derived from first principles and tested in the field

This process benefits not just product development but Lean manufacturing, Agile, and Kanban, too.

Lean Startup

Eric Reis's Lean Startup model takes product development into areas of extreme uncertainty, where basic things like customers, business models, and even the basic shape of the product are largely unknown. Here, failures are hard to avoid, but they should be painful, not catastrophic (I have in mind the widespread economic damage inflicted by the dot-com crash of the late '90s). Like Lean manufacturing, Lean Startup has two principal approaches to change.

Its continuous, incremental mode (analogous to *kaizen*) is organized around an experimental improvement loop called *Build/Measure/Learn* that bears strong similarities with PDCA:

♦ **Build** the smallest possible product increment that can test a hypothesis.

♦ **Measure** the real-world impact of this increment.

♦ **Learn** from the results, changing assumptions or building on them.

This approach is highly suited to web-based services, where techniques such as *continuous delivery* and *A/B testing* allow products to evolve very rapidly.

When continuous evolution runs out of steam, something more radical is needed. Lean startup's *kaikaku* is the *pivot*. This rather abused term often gets taken to mean "throwing the original idea away and starting again," but it is meant to indicate a significant but disciplined "course correction" that results from the invalidation of specific fundamental assumptions.

My special interest in Lean Startup and the allied discipline of *Lean UX* (the "UX" standing for *User Experience*) stems from what they do to feedback loops. This isn't just a philosophical point—from current first-hand experience inside one of the UK government's "digital exemplars,"[49] I know that you don't need to be working for a startup business to benefit from this kind of approach.

Recall from the previous chapter how XP moved the focus of feedback loops "rightward," so that they are centered not on requirements or design but on the code. Lean Startup adds feedback loops whose focus is even further to the right, keeping under constant examination the interactions between the product and its customers, looking to validate (or otherwise) the assumptions about customers and their needs that drive the development of the product.

Lean Startup continues a trend away from *requirements*, the basic raw material of traditional projects. XP introduced *user stories*, "a placeholder for a conversation." Lean Startup replaces these with *hypotheses* that need to be validated. When you're that committed to putting your assumptions to the test, your relationship to work-in-progress changes quite fundamentally.

Lean/Agile Hybrids

There's no doubt that Lean has much to teach Agile, and vice versa. But things can get a bit clunky when one has to keep saying, "I get X from Lean, and Y from Agile." So why not create a hybrid of the two, combining Agile practices with Lean's scale, say?

This is not without risk. In the absence of the values, Agile practices are hardly Agile at all. Lean tools without Lean thinking? Same problem. And the bigger and more rigid the hybrid's process, the less sure we can be that it will suit the organization into which it is introduced.

As I write, it is far too early to tell whether off-the-shelf Lean/Agile hybrids (of which SAFe is the best-known example) will achieve anything

---

49. See https://www.gov.uk/transformation

like the cultural impact of their antecedents. In the meantime, I would much rather give Lean and Agile the separate credit they each fully deserve, hoping that their intelligent and respectful application in context is what is generally meant by "Lean/Agile." I would hope to see a healthy dose of Kanban in there, too, of course!

Kanban and Lean

Let's turn the tables and look at the Kanban Method from Lean's perspective.

Kanban has, quite explicitly:

◆ Visual pull systems, represented in Part I by the values of **transparency** and **balance**

◆ Respect for people (including much of what goes with that concept inside Toyota) in the form of **collaboration, leadership, understanding, agreement,** and **respect**

Kanban doesn't incorporate Toyota's True North, leaving each organization to determine its own fitness criteria. However, the values of **customer focus, flow,** and **respect** map to Toyota's True North very well, and they each suggest some good things to measure.

From Lean's perspective, Kanban could be described as the fruit of the following process:

1. Taking two of Lean's tools—*kanban* and *heijunka*—and radically "reimagining" them for use in creative knowledge work
2. Observing that the implementation of these tools can have organizational impact
3. Identifying from successful implementations a set of principles and practices
4. Growing a new body of knowledge around those principles and practices

This version of history rather ignores the roles played by TOC and Agile in Kanban's development, but still, it is the case that Kanban is highly aligned with Lean, is built on part of it, and, through the practice of *using models,* continues to refer back to it with some enthusiasm.

As a method and as a community, Kanban maintains some distinctiveness, however:

◆ We tend not to adopt Lean jargon, especially—and with the irony duly noted—the Japanese terms. There is enormous respect for and interest in the Lean and TPS heritage, but we take some effort to use plain language where possible.

◆ We are uncomfortable with the use of manufacturing as a metaphor for creative knowledge work. Neither do we wish to be remembered mainly for *waste* and *kaizen*; important as these are, creative knowledge work is as much about maximizing future opportunity as it is about optimizing current process.

All that said, we're still far from done with Lean. The next chapter owes much to Don Reinertsen; Part III (Implementation) has Lean influences also.

Further Reading

Middleton, Peter and James Sutton. 2005. *Lean Software Strategies: Proven Techniques for Managers and Developers.* New York: Productivity Press.

Ohno, Taiichi. 1998. *Toyota Production System: Beyond Large-Scale Production.* New York: Productivity Press.

Poppendieck, Mary and Tom Poppendieck. 2006. *Implementing Lean Software Development: From Concept to Cash.* Upper Saddle River, NJ: Addison-Wesley.

Reinertsen, Donald G. 2009. *The Principles of Product Development Flow: Second Generation Lean Product Development.* Redondo Beach, CA: Celeritas.

Ries, Eric. 2011. *The Lean Startup.* New York: Crown Business.

Rother, Mike. 2009. *Toyota Kata: Managing People for Improvement, Adaptiveness and Superior Results.* New York: McGraw-Hill.

Spear, Steven J. 2009. *The High-Velocity Edge: How Market Leaders Leverage Operational Excellence to Beat the Competition,* 2nd ed. New York: McGraw-Hill.

❖ Chapter 15 ❖

Economic Approaches
to Flow

This chapter presents some economic concepts that are important in Kanban:

♦ *Cost of delay* (CoD) is associated with Don Reinertsen and Lean Product Development, but it is very relevant to service operations also.

♦ *Cost of Carry* is a tool used in many traditional industries, in retail, and in banking (my background).

♦ *Real Options* is being popularized in the Kanban and Agile communities by Chris Matts and Olav Maasen. In passing, we draw on some elements of *information theory*, as described by Douglas Hubbard.

These new tools don't necessitate deep economic quantitative-analysis skills. Such an approach would scale very poorly as process evolves in the direction of **flow** (Chapter 5) and the units of work get progressively smaller. Rather, they help us evolve principles and policies that guide fast and effective decision making—in Don's words, a *decision framework*.

But before we look at those, we first examine the economics of product development's traditional delivery vehicle, the *project*. We'll see that under a set of assumptions that are representative of a broad range of product development projects, project economics are very much weaker than they first appear.

ROI and the Pareto Diet[50]

So near yet so far! Your project seemed close to receiving approval, but another round of serious revision has been called for. Your current plan shows that you will keep a dedicated team of people occupied for a year. Delivery will be at the end of that year's work, yielding an estimated return on investment (ROI) of 25%.

It has been suggested that you look at reducing the project's overall scope and cost, trimming features and realizing benefits earlier. There's a tradeoff, though: The project's fixed costs will make up a larger proportion of the whole. You agree to work out what these changes will do to the project's ROI.

For the sake of this exercise, two parameters are assumed:

1. Those *fixed costs* (costs that don't vary at all with project scope) are a hefty 10% of the current total. In effort terms, that's more than a month's worth, even if the revised project were to deliver only the smallest possible amount of change.

2. 60% of the value comes in the first 40% of time and effort. Breaking that down further, returns at feature level form a Pareto distribution (a rather flat one relative to the 80–20 rule associated with Pareto).[51]

Also, we must make explicit a property common to many projects:

3. Outside of a small core, features can be dropped without fatally undermining the logic of the project.

As a function of project scope, the value that can be delivered by this project is represented in Figure 15.1. The full project is believed to be worth 125% of what it costs to deliver.

As we reduce the scope of the project, fewer of the lower-value items will be delivered (rationally, we would give the higher-value items priority), so the line gets less steep with larger scope.

Let's add a line for the cumulative cost, as shown in Figure 15.2. Because of the fixed costs, this starts above the origin; naturally, it grows steadily with time thereafter. At 100% of original planned scope (achieved after one year of work), the gap between value and cost corresponds to the project's 25% ROI.

---

50. I owe "Pareto Diet" to Apple. I intended to tweet "Pareto dist" (for "Pareto distribution") and was gifted this very apt autocorrection.

51. See http://en.wikipedia.org/wiki/Pareto_distribution

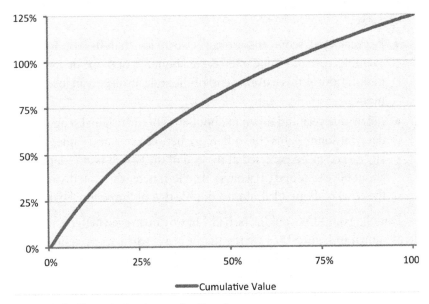

Figure 15.1   The project's value as a function of scope

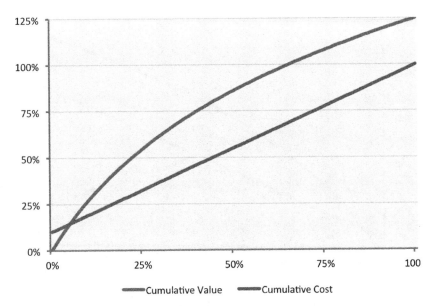

Figure 15.2   Value and cost as functions of scope

Observe:

♦ For very small scope, the project is worth less than its cost. In this example, cost and value intersect at about 5% of scope or 18 days in duration; at this extreme end of the scale, projects can lose money.

♦ Below one year and down to about 30% of the original scope (109 days), absolute gains (the difference between cost and value) actually exceed those expected at the end of the year. It is at its largest at about 60% of scope (219 days). The ROI here (the gain divided by the cost) is about 49%, almost double that of the original plan.

Doubling our ROI wouldn't be bad, but we can do even better. Graphing it, as shown in Figure 15.3, we see that it peaks when scope is limited to about 26% of the original (95 days), where the project returns 68%.

This seems a little absurd! Is it credible?

Quantitatively, I would take even the original 25% ROI figure with a pinch of salt; it's the ratio of two uncertain numbers. Would I then take the 68% seriously? Let's just say that I wouldn't bet my career on it.

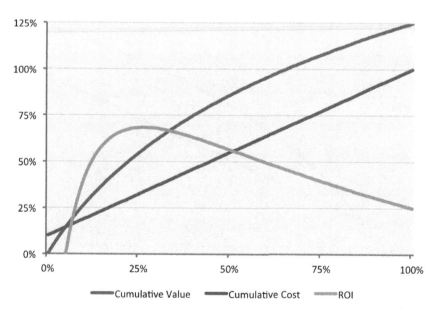

Figure 15.3   ROI peaks when scope is limited

Qualitatively, though, it's robust enough. You can change the original ROI or duration, the shape of the value distribution, or the fixed cost, and the same conclusion will be reached: If you have flexibility over scope, slashing it is likely to improve returns. Moreover, the choice of parameters is rather conservative; the effect is even stronger if the fixed costs aren't so high or if the shape of cumulative value curve is more pronounced.

Furthermore, this ROI-based treatment actually understates the true cost of long projects. Consider:

- Some features may bring additional benefit if delivered sooner (we've made the often poor assumption that feature-level ROI is completely time-independent). Delayed delivery incurs a kind of *opportunity cost* we refer to as *cost of delay*.

- Like any financial structure that involves making up-front commitments in exchange for future returns, projects have a *cost of carry*; projects represent *inventory* that must be financed somehow. Moreover, the money, people, and other assets tied up by the project might deliver better returns if allocated elsewhere (*opportunity costs* again).

- As time passes and the world changes around the project, more work must be done to keep everything properly aligned. In this sense, projects are *wasting assets*, perishable goods.

- The early implementation of some features might yield the intelligence that they or other planned features should have been done differently or not at all. There is a cost, therefore, to making early scope commitments. Equivalently, there is a cost to precluding the opportunity for feedback, and there is value in keeping options open.

Even before we explore these hidden costs, it should be clear that this challenges the very notion of a *project*. Under conditions of scope flexibility, in which the timing of individual features may be significant, or when there are negative consequences to premature commitment, what we're really talking about is a mere aggregation, not some irreducible whole.

In more appropriate language, it's a *batch*; and in this case, a batch that is far too big. Perhaps a coherent business case can be constructed around the batch as a whole; perhaps there's a gnarly, dependency-ridden core that fully deserves to receive skilled project management; even so, there really is something a little absurd about treating the whole endeavor that same way.

Using the language of batches, your ideal batch size[52] may be much smaller than you think. It is likely that you can afford to reduce your batch sizes significantly without waiting for reductions in *transaction costs* (the fixed costs per batch) to come through first. In fact, smaller batches will make it much easier to see those transaction costs for what they really are. Made more visible, they will inevitably become a focus for improvement. And in the meantime, better returns!

This is a very helpful way of looking not just at the individual project, but at the whole portfolio. We see that conversations about scope are conversations about batch sizes; these are things we can and should make visible (**transparency**). Reductions in batch sizes (perhaps policy-driven) are reductions in WIP, with consequences for **balance** and **flow**. At stake here are huge benefits to customers, workers, and the organization that will never be obtained from focusing on project execution alone.

Cost of Delay

Cost of delay (CoD) is an elegant way to understand the time-dependence of value and a good guide to scheduling decisions. I first encountered *cost of delay* through Don Reinertsen's *Principles of Product Development Flow* and David Anderson's integration of the concept into Kanban's *classes of service* (see Chapter 2).

My treatment of cost of delay begins not in terms of quantification, but in those of recognition. The language used to describe the urgency of a work item tells us much more about cost of delay than sometimes we realize. The following examples highlight not technical terms, but language signals (with some hidden messages):

1. "Our disk space situation is becoming *increasingly urgent*" (with each day of delay comes a significantly increased likelihood of an outage).

2. "We're out of the market—we need a fix *now!*" (Each hour of delay is costing us hundreds of thousands of dollars, hugely disproportionate to the actual cost of the fixing the disk space issue.)

---

52. I'm being careful here not to use the technical term *economic batch size*. This is typically calculated by a formula that takes only some of our concerns into account; it is a useful guide nevertheless.

3. "*Next Friday* you will explain to the board how this disaster happened" (not just dollars, but reputations are at stake; to go into this meeting unprepared would be foolish).

4. Some time earlier: "We should think about starting that disk monitoring project sometime" (it seems important somehow, but how urgent is it?).

These examples reveal different kinds of urgency—they're qualitatively different. In fact, these language signals represent patterns so common that we've given them names already:

1. "Increasingly urgent," "worth $X per year," "recurring savings," "compound growth," "competitive advantage," "a priority"—this is the language of *standard* work items. Here, the relative urgency between work items is key to the decision making process.

2. "Now," "right away," "drop everything"—this is the language of the *expedite* work item type. It's important that we recognize these quickly; there's no time for cost/benefit analysis here! Retrospectively, though, we look back and learn (there is an example in Chapter 5 of how we managed to reduce our support workload).

3. "Next Friday," "quarter end," "seasonal opportunity"—of course we're talking dates here. In those genuine cases in which early delivery has only a small benefit relative to the heavy penalty of late delivery, we have *date-driven* (or *fixed date*) work. The challenges here are in deciding whether the work is worth doing at all, to start it at an appropriate time, and to keep schedule risk under proper control until the work is done.

4. Lastly, the tricky category, for which it's almost unreasonable to describe exactly how urgent the work is; these are the *intangible* work items. These can be the small improvements we like to keep ticking over (in aggregate, they're vital), the market experiments that may have a massive payoff or none, or the risk mitigations that may or may not save the day.

In theory, the quantitative side of cost of delay is equally applicable to all of these work item types. Practitioners, however, are smart enough to use the work item type to save some unnecessary work. They use it mainly for two things:

1. Examining the rationale for those fixed date items. Once selected, the assumption is that they will be delivered on time; the problem then becomes one of risk management.
2. The sequencing of those standard items

Across those two types, we choose and sequence work items in such a way that cost of delay is effectively minimized. The other two types are effectively treated as overhead; what's relevant here isn't so much the specific items, more their overall weight in proportion to the whole.

In economic terms, we seek to minimize *opportunity cost*. Finding a theoretically optimal delivery sequence is very hard work, but again, the practitioner is smart enough not to try. Long lists of work items will take significant time to deliver; new work will arrive and priorities will change in the meanwhile. Optimal sequences aren't very useful when the goalposts keep moving! What's needed is a workable strategy, a *heuristic*.

One simple and very effective heuristic is to seek to reduce the cost of delay as quickly as possible over the shortest possible time. Over the near term, this is achieved by giving priority to the items that have the highest cost of delay divided by their expected duration. The shipping company Maersk gives *cost of delay divided by duration* (a financial throughput calculation) the cute mnemonic "CD3"; the resulting *queuing discipline* (the decision rules by which we manage the queue of work) is known as *weighted shortest job first* (WSJF).

In executing that queuing discipline, pairwise comparisons between work items look like this:

◆ Do the two items have a similar cost of delay? Choose the one that can be delivered the soonest.

◆ Are they comparable in (remaining) delivery time? Choose the one with the highest cost of delay.

◆ If there is no obvious winner by the above rules, compare CD3 ratios.

◆ You might consider upside and downside risks if a tiebreaker is needed. Consider the chances of runaway success, spectacular failure, or extended delays to delivery. Marginal differences hardly matter here; toss a coin if you need to.

These decision rules work whether we have a detailed financial model for our product (where the cost of delay is measured in the loss of *life cycle profits*, for example), or simply T-shirt sizes (SML for small, medium, or large cost of delay, and short, medium, or long duration; M/S beats S/M, for example). Bear in mind, though, that there may be order-of-magnitude differences in cost of delay at the large end. Don't make the mistake of assuming that large items aren't worth the trouble.

Using an intermediate level of sophistication, here's an illustrative list of work items sequenced appropriately:

◆ Item 1: 10 days duration, it is worth up to (we think) a 5% increase in annual sales if we can steal a march on the competition. Days count.

◆ Item 2: 10 days duration, it is worth a 7% increase in annual sales.

◆ Item 3 (the cheapest): 20 days duration, it is worth a 14% increase in annual sales.

Despite its low absolute value, the decision to rank item 1 first is justified on the grounds that its cost of delay is highest. A 10- or 20-day delay could result in the opportunity being lost forever; the others can wait that long.

The decision to rank item 2 ahead of item 3 is more marginal. From the limited information available, we conclude that they reduce cost of delay at about the same rate. Experience tells us that the longer-duration item is significantly more likely to suffer unexpected delay, so we make a risk-based decision.

You may be surprised that cost was not a factor in any of these decisions. We will do all three items soon enough and exactly the same amount of money will be spent overall. A lot of things become a whole lot easier once a high-level budget is in place and the decision-making process is free to focus on the results that can be obtained.

The Cost of Queues

Cost of delay has another very useful application. If you know

◆ the amount of time typically lost to queuing in your system,

◆ the average cost of delay per work item, and

◆ the number of items queued on average,

a quick multiplication puts a price on those queues. From that basis you can examine the economics of an investment in capacity (adding people, say) or a process improvement to reduce those queues.

Cost of Carry

Suppose an investor has been given the unique and exciting opportunity to finance your organization's projects. An interesting deal has been struck, one that aligns the interests of investor and sponsor:

♦ The investor provides money in the form of a loan at the start of each project, covering the expected cost of the work.

♦ At the project's due date, the loan becomes repayable with interest. However, the amount actually repaid depends on the project's success:

　♦ If completed on time with its anticipated benefits realizable as planned, the entire amount outstanding (principal plus interest) is returned to the investor in full.

　♦ Late or otherwise disappointing outcomes repay significantly less, reflecting additional financing costs and uncertainty in future prospects.

　♦ Failed projects return nothing at all; the investment is lost forever.

Put yourself in our hypothetical investor's shoes. How much interest would you require for such an investment to be worthwhile?

Look at any typical corporate project portfolio and it soon becomes apparent that these investments can be viable to the investor only if they carry very high rates of interest. Your credit card looks cheap by comparison! In the credit markets, high rates of interest imply not high quality but the reverse—we're in the realm of junk bonds and distressed debt, rates of 30%, 40%, and more. [53, 54]

Factors driving any calculations involving these investments will include:

♦ Project (and loan) duration—longer projects must pay more interest

♦ The likelihood of project success or failure—lower yields (so less interest) where outcomes seem likely to be good, higher yields (and more interest) on riskier projects

---

53. Simplistically, for the investor to receive an effective rate of r_i for a one-year loan with an expected failure rate of f, the rate charged r_c is given by

$$r_c = (1 + r_i) / (1 - f) - 1.$$

For a not-untypical failure rate of 25% and an effective rate to the apparently risk-hungry investor of just 5% (which leaves little if any room for profit after financing costs), we get an r_c of 40%.

54. Contrast those rates with the 5% prescribed by the UK Treasury. This says to me that that money is considered to be free and that economically, project duration and risk hardly matter. Perhaps public projects don't actually need to succeed—it's enough that they generate economic activity; by the time they fail, their sponsors will be long gone.

- The ability to salvage something of value from the near failures
- The investor's own financing costs (money isn't free)

These factors don't work independently of each other. Project duration in particular has a very strong impact on the others. Long projects are disproportionately more risky and expensive, and when they fail, the chances of recovering anything of value is diminished. In our thought experiment, long projects should expect to pay not just more interest, but higher rates of interest. High rates and compounding over long periods combine to make the overall effect decidedly (and expensively) non-linear.

Cost of carry measures the cost to the organization of the inventory that it holds. At its most basic, it is a function of lead times and overall project cost, perhaps using fixed rates specified by the finance function. Even at this low level of sophistication, cost of carry provides a simple way for the practitioner to quantify some of the financial benefit of portfolio-level improvements, and it can be applied even where cost of delay is hard to determine.

More sophisticated thinking comes from combining these concepts and using realistically large interest rates. Is the supposed cost of delay worth the risk-adjusted cost of carry? This is how I justify my strong bias toward work items of shorter duration—it's not that longer duration work items should never take priority, but they must represent a disproportionately valuable opportunity.

Options

Real options takes from the world of banking the idea of an option instrument (an important kind of financial derivative) and applies it in the field of project evaluation. Where a project's business case depends on an observable market price, changes in either that price or its volatility affect the viability of the project. This mirrors the pricing of traditional financial options, which depends on the price of some underlying security, its volatility, and the duration of the option, together with interest rates and other external factors.

Good examples of the large-scale application of real options come from the energy sector. The viability of projects to build oil-fired power stations or to develop oil reserves will be highly dependent on oil prices. High prices will make the power station seem a poor investment (the option to build one will be less valuable) whereas oil reserves will appear more attractive (the option to develop them will be more valuable to hold or purchase). Interestingly, both options increase in value as the volatility of energy prices increases; even when current prices seem unattractive, high volatility increases the likelihood that viable conditions will emerge during the life span of the option.

In both of these examples, companies purchase the options to implement these projects later. They buy licenses and land; they make agreements in principle with governments, regulators, and suppliers, both upstream and downstream. All of these actions cost real money, but none of them necessarily obliges the company to build anything or to extract oil from the ground. They have bought the right without the obligation; they have deferred commitment. Contractually or practically, though, these options don't last forever—agreements in principle don't last indefinitely and licenses have explicit termination dates.

Most of us don't have the luxury of an easily observable market price by which to estimate the value of our options. But all is not lost! Qualitatively, options thinking is highly applicable to creative knowledge work, and even some of the quantitative aspects translate surprisingly well.

Chris Matts and Olav Maasen have distilled options thinking into three principles:

◆ Options have value.

◆ Options expire.

◆ Never commit early unless you know why.

Options Have Value

Philosophically, this is self-evident. Economically, what's particularly interesting is the value of the information that some options can generate. For example:

> Project X is considered highly risky. It will cost $100,000. It may generate $1,100,000, but based on our current market knowledge we attach a probability to this of only 10%. We therefore calculate its expected value to be:
>
> $$-\$100,000 + 10\% \times \$1,100,000 = \$10,000.$$
>
> Project Y consists of $5,000 of market research that could tell us for sure whether project X should go ahead. How much is

this combination actually worth? Assuming we will act rationally on our perfect knowledge:

$$-\$5,000 + 10\% \times (\$1,100,000 - \$100,000) = \$95,000$$

Quite a difference, a nearly tenfold improvement!

This is a contrived example and we may not often do this kind of math, but it is still very smart to look for cheap ways to answer questions such as *"Can we do it?" "Will they like it?"* and *"Do we really need it?"* And when the cost of delay is high enough, it can be well worth pursing multiple lines of attack in parallel so that an acceptable solution will be found in the shortest possible time.

Options Expire

When does the option to deliver a new Christmas product expire? Long before December 25th, that's for sure! In two words, we are reminded that date-driven work means more than just meeting deadlines; it's about maximizing opportunity. This means starting at the right time or not at all—and when the window of opportunity is short and the profit opportunity is not sufficiently large, the right course of action may be to let it pass.

Sometimes we will invest in these options in the full knowledge that they will soon expire. It costs very little to register a web domain, for example; you are still in control when the time to renew the registration comes, and no sleep will be lost if the opportunity to use it passes.

Never Commit Early Unless You Know Why

This looks like another truism, but it is well worth repeating. How different would most projects look if for every planned feature someone asked this question:

What would have to be true for this option to look fantastic?[55]

Where the answers to this question are unknown, a new layer of exploratory options can be generated.

Typically, the project provides the perfect mechanism for avoiding this question. Scope, duration, and cost are all decided before the questions are

---

55. Roger L. Martin, in Lafley, A. G. and Roger L. Martin. 2013. *Playing to Win, How Strategy Really Works*. Boston: Harvard Business Press.

even asked, let alone answered. By design, changes to any of these cause drama and stress!

Contrast that to an options-based approach. Instead of a predetermined project backlog, we have a portfolio of options, a growing pool of uncommitted ideas that may or may not come to fruition. Options get exercised—that is, work gets pulled—when they will generate the most valuable information relative to all their alternatives.

Options thinking also does something interesting to risk management. Outside the development process, uncommitted options remain in the hands of those best placed to manage them, for example, those with the market knowledge to maximize opportunity. Once inside the system, though, the risks change in both nature and ownership—there are expectations to meet. There is more than meets the eye in the act of pulling of work across these boundaries—it creates obligation and transfers risk. It is not to be done lightly.

Putting It All Together

◆ Put your project portfolio on a diet—aggressively reduce batch sizes toward the likely current ideal.

◆ Don't treat all work as alike. At all levels, understand and classify work by its urgency profile; control the overall mix of work.

◆ Within each urgency classification, learn to identify work that carries a high cost of delay or high information value. Implement queuing disciplines that maximize this throughput.

◆ Make your options visible; if you're not sure that they're "*fantastic*," generate additional exploratory options instead of pulling them prematurely.

◆ Whether through cost of delay or cost of carry, understand the cost of your WIP. At each level, find ways to control it, and expect it to keep on reducing.

◆ Continually seek to reduce transaction costs, but don't make this a prerequisite for any of the previous steps. You *can* afford to go faster!

Further Reading

Arnold, Joshua J. and Öslem Yüce. 2013. *Black Swan Farming Using Cost of Delay.* http://blackswanfarming.com/experience-report-maersk-line/

Hubbard, Douglas W. 2010. *How to Measure Anything: Finding the Value of Intangibles in Business,* 2nd ed. Hoboken, NJ: Wiley.

Maassen, Olav, Chris Matts, and Chris Geary. 2013. *Commitment.* Amsterdam: Hathaway te Brake Publications.

Reinertsen, Donald G. 2009. *The Principles of Product Development Flow: Second Generation Lean Product Development.* Redondo Beach, CA: Celeritas.

❖ Chapter 16 ❖

The Kanban Method

The preceding 15 chapters show the Kanban Method from a number of different perspectives. Part I is an unusual inside view, centered on Kanban's values system. Thus far, Part II has explored the continued relationships that Kanban has with other bodies of knowledge, namely Systems Thinking, Theory of Constraints, Agile, and Lean.

With a sprinkling of personal perspective, this chapter organizes for reference the key elements and resources of the Kanban Method and its community.

A Very Brief Timeline

The early years look like this:

2004: Working at Microsoft, David J. Anderson and Dragos Dumitriu collaborate on what was to become the XIT case study. This was later described in Chapter 4 of the "blue book" and referenced in Chapters 8 and 12 of this book.

2005: With the input of Don Reinertsen, XIT is recast as a kanban system, with only minor changes.

2006: David moves on to Corbis (the Microsoft-owned media rights company) and continues the development of the method there.

2007: The Kanban Method made its first public splash at a fringe meeting of the Agile 2007 conference. Strong interest from delegates from around the world results in a Yahoo-based online discussion group *kanbandev*[56] being created (Yahoo is among Kanban's early adopters).

2009:   *Lean & Kanban 2009*, the Kanban community's first conference, is held in Miami, Florida. This spawns the *Limited WIP Society*[57] an informal umbrella for local meet-up groups and other community events and resources.

2010:   David publishes the "blue book," *Kanban: Successful Evolutionary Change for Your Technology Business.*

2011:   In Reykjavik, the first *Kanban Leadership Retreat*, a small *unconference*, is held for experienced practitioners, community leaders, and business partners. Under the umbrella of Lean Kanban University (LKU),[58] the retreat spawned what would become the multi-supplier Accredited Kanban Training (AKT) program and a program for individual coaches called Kanban Coaching Professional (KCP). (I was there at the Reykjavik retreat, but being in neither the training nor the coaching business at the time, I stayed away from those meetings. However, I soon got involved, taking responsibility for LKU's curriculum, among other things.)

Fittingly for an evolutionary method, things have continued to develop from there. The Method's definition has been through multiple revisions. At the time of this writing, annual conferences are held in eight locations worldwide. The leadership retreats continue once or twice a year, and kanbandev has grown to more than 2,500 members.

In 2012, David published a retrospective on this period in the form of a curated collection of blog posts called *Lessons in Agile Management: On the Road to Kanban*. It is surprisingly substantial, comprising upwards of 150 articles from 12 years of David's Agile Management blog.

Lessons's oldest posts were written before the publication of David's first book, *Agile Management for Software Engineering: Applying the Theory of Constraints for Business Results*. David himself describes this first attempt as more theoretical than practical. Don't tell anyone, but I haven't gotten around to reading it yet.

---

56. https://groups.yahoo.com/neo/groups/kanbandev/info
57. http://limitedwipsociety.ning.com/
58. http://edu.leankanban.com/

Foundational Principles

These encapsulate the Kanban Method's philosophy with regard to change:

FP1: Start with what you do now.

FP2: Agree to pursue evolutionary change.

FP3: Initially, respect current processes, roles, responsibilities, and job titles.

FP4: Encourage acts of leadership at every level in your organization —from individual contributor to senior management.

The wording of FP2 has been simplified slightly since 2010, removing the word "incremental."

Collectively, the first three foundational principles describe a distinctive evolutionary approach, the open-ended pursuit of adaptability and fitness for purpose. No less distinctive is its advice to avoid confronting too early the "rocks" of roles, responsibilities, and job titles.

FP4 was added after the blue book, and sometimes we omit it, not because we don't like it but because it is not so explicitly about evolutionary change. All four are strongly related, however: I group the values that correspond to first three principles—**understanding, agreement**, and **respect** (Chapters 7, 8, and 9)—under the label *leadership disciplines*, clearly tying them to the fourth, **leadership** (Chapter 6).

Core Practices

The Foundational Principles describe how to approach change; the Core Practices describe how to keep on provoking it:

CP1: Visualize.

CP2: Limit Work-in-Progress (WIP).

CP3: Manage flow.

CP4: Make policies explicit.

CP5: Implement feedback loops.

CP6: Improve collaboratively, evolve experimentally (using models and the scientific method).

We often read "with or around kanban systems" into these practices, but there are times when it is helpful to make the most of the broader interpretations that this standardized wording allows. There are examples of that later in the chapter.

Chapter 1 groups CP1, CP4, and CP5 together under the single value of **transparency**. CP2 and CP6 correspond to **balance** and **collaboration** (Chapters 2 and 3, respectively). CP3 expands into two values, **customer focus** and **flow** (Chapters 4 and 5; also Chapter 15).

CP1 and CP4 both have had their wording simplified since 2010. CP5 is a later addition.

Some literature (older material, mostly) refers to the Core Practices as the "Core Properties" instead. This label makes good sense when referring specifically to how they were first recognized in the wild, but it seems somewhat arcane otherwise and is best avoided.

Contextualized Kanban

Some applications of Kanban are so common that they have their own names. Presented here are three of them: *Personal Kanban, Portfolio Kanban,* and *Scrumban.*

Personal Kanban

In *Personal Kanban: Mapping Work | Navigating Life,* Jim Benson and Tonianne DeMaria Barry describe how Kanban can be applied to one's personal workload.

Jim and Tonianne distill Kanban down to the two practices most relevant to *"choosing the right work at the right time"*:

1. Visualize your work.
2. Limit your work in progress.

I'm a big fan of the book; three generations of my family have read it! Although it identifies only two practices, and it does not need to concern itself with organizational impact, it explores **transparency, balance,** and **flow** pretty deeply (not that it explicitly describes these as values).

Add **collaboration** and you have the beginnings of what I hesitate to call "Team Kanban." This is five values short of the nine, but in all fairness there is still plenty to write about here. Out of Sweden (as countries go, an early adopter of Kanban) comes a good introduction by Marcus

Hammarberg and Joakim Sunden called *Kanban in Action*. If you're in a team that is getting started with Kanban, this book is worth the cover price for the chapter "Kanban Pitfalls" alone.

Portfolio Kanban

By design, the Kanban Method is open to a range of interpretations and imaginative applications. As Chapter 2 suggests, preceding the core practices with "*Find ways to . . .*" is a good way to encourage people to think more creatively about the organizational problems they can begin to address.

Let's try that with the project portfolio. *Start with what you do now,* and

◆ Find ways to organize the project portfolio visually. Slice by business initiative, by customer, by team; at the granularity of whole projects, releases, features, or experiments.

◆ Find ways to limit work in progress, from team level upward and from portfolio level downward. Limit batch sizes by features, time, and money. Stop starting and start finishing!

◆ Find ways to manage the portfolio for smoothness and timeliness. Understand the cost of delay on active and upcoming work, developing the discipline to prioritize accordingly. Manage product lines and services end-to-end, funding and staffing them accordingly.

◆ Find ways to capture, share, devolve, and evolve the portfolio's decision framework.

◆ Find ways for decision makers, teams, and customers to participate so that signals will be seen and acted upon.

◆ Find ways to bring different parts of the organization together to tackle common problems. Share successes and build on them.

Portfolio Kanban is the application of this kind of thinking, not just the use of a one-project-per-sticky kind of visualization. Because organizations differ widely and changes to portfolio management policies tend to take weeks or months to work their way through the system, actual practice at any given time tends to be highly contextualized.

Part III of this book is applicable to portfolio management but not specific to it. I'm unable to recommend any other books here, but you will find some relevant material on my blog, on Pawel Brodzinski's, and on Ian Carroll's. See the Further Reading section at the end of this chapter for details.

Scrumban

Scrumban is a name coined by Corey Ladas, for what happens when *what you do now* is Scrum and you apply Kanban.

I stress again the cautions of Chapter 13: Kanban does not mean recklessly throwing out all of your Agile discipline; rather it's a transformative process that takes time, thought, care, and collaboration.

This progression is typical:

♦ Already practicing a degree of visualization, the team organizes work according to its "done-ness." This extends beyond "code complete," "demo-able," or "potentially shippable" to cover acceptance, deployment, and customer validation states.

♦ Increasingly, standup meetings are organized around the board.

♦ Already limiting work-in-progress through the sprint mechanism, the team pays more attention to the amount of work started but not yet finished. As a result, they start to see work items getting completed sooner. Immediately or after seeing the board operating well, explicit WIP limits may be introduced.

♦ With work items completed sooner and more visibly, greater attention is given to the later stages of the process. Impediments to continuous delivery start being addressed. The nature of the sprint begins to change as releases are planned independently (if they need much planning at all).

♦ Having decoupled releases from sprint planning, the system now easily accommodates work of different types and speeds. The team pays attention to the cost of delay of individual items and to the mix of work overall. Mid-sprint changes become much easier to accommodate; classes of service may be offered.

♦ The rhythm of sprint planning continues, but the meeting itself gets easier. Estimating the right amount of work for the sprint seems less important; it's enough to ensure that there is sufficient work of high enough value and quality and that the riskiest items have been identified and broken down where necessary.

◆ With the need for customer validation made more visible, new feedback loops begin to emerge.

Different attitudes toward the Scrum practices and roles will of course lead to different outcomes. It's not unusual for teams to go through changes of the kind described here and for them still to identify themselves with Scrum. That's completely fine with us—it's "Kanban with," not "Kanban versus."

The team I'm currently working with as its interim development manager is a long way down this path. The planning rhythm is still there and I'm in no hurry to see it disappear. The highlight of my working fortnight is the "Show and Tell" (the sprint review), a lively meeting in which the project team is outnumbered by customer representatives and outside observers (as a "digital exemplar," one of a number of pioneering citizen-facing projects delivering services online for UK Government departments, we often receive visitors from other departments and public agencies interested in how we do things). Not only do we review progress and show what we've recently built, we often review pertinent videos of outside customers interacting with the live system or with prototypes. These are highly motivating—sometimes even moving—and the shared experience adds to its impact.

Enabling Concepts and Tools

These may or may not be considered first-class components of the Kanban Method (to me they are) but they're certainly important and helpful enough to be worth identifying here.

Topics covered elsewhere in this book:

◆ The nine values (Chapters 1–9)
◆ The three agendas (Chapter 10)
◆ The Kanban Lens (Chapter 10)

Full references to these are at the end of this chapter:

◆ The *Kanban Katas*, originated by Håkan Forss and inspired by Mike Rother's *Toyota Kata*. These provide some patterns for key feedback loops.

- The *Kanban Depth Assessment* tool (a.k.a. *"How deep is your Kanban?"*), the product of the 2012 Kanban Leadership Retreat. I have a complicated relationship with this tool:
  - I was in the room when it was created (the retreats are a fantastic source of moments like that). We tried to answer Håkan's question *"Are the practices in the right order?"* and David's *"Are we doing Kanban or not?"* We came out with the conclusion that a much better question is *"How deep?"*
  - The values of Part I were in part provoked by my regret that we had focused on Kanban's core practices at the expense of the foundational principles.[59]
  - Putting my philosophical objections to one side, I find the tool very helpful in practice. An alternative, values-based version of it is covered in Chapter 23.
- Kanban's *Flight Levels*, originated by Klaus Leopold. This describes patterns of scale rather than of depth. Together with the values and the lens, Klaus's model helped to catalyze the agendas.

Implementation Guidance: STATIK

Methods often come without much in the way of guidance on how they should initially be introduced. You might think that the *start with what you do now* method might manage without implementation guidance, but we do in fact have the *Systems Thinking Approach to Introducing Kanban*, otherwise known as *STATIK*.

Chapter 7's expansion of the first foundational principle borrows from STATIK:

FP1 (expanded): Start with what you do now, understanding

- The purpose of the system
- How it serves the customer
- How it works for those inside the system

---

59. I am reminded of Keith Sawyer's book *Group Genius: The Creative Power of Collaboration* (2008. New York: Basic Books), which holds that all creative work is collaborative, even when the appearance is of solo effort. Small wonder that our global community invests so heavily in meeting face-to-face. I thank my friend Markus Andrezak for that reading recommendation.

♦ How it leaves customers dissatisfied and workers frustrated

♦ How it can be changed safely

I believe STATIK should indeed be regarded as a first-class component of the Kanban Method—as important as any of the others covered in previous chapters—and I regret that it is not more widely known. I hope that Part III of this book does something to address that.

At a high level, STATIK's six steps are as follows:

1. Understand sources of dissatisfaction.

2. Analyze demand and capability.

3. Model workflow.

4. Discover classes of service.

5. Design kanban systems.

6. Roll out.

This is the basic version, as taught in a regular two-day class. But just as POOGI-0 (Chapter 12) properly anchors the basic POOGI with a step 0, *STATIK-0* includes this:

0. Understand the purpose of the system.

With or without that step zero, STATIK is a good way to frame post-rollout change also. It reinforces the discipline to base change on meaningful needs rather than on mainly technical considerations (board design, WIP limits, and so on) for their own sake.

We are beginning to document "Reverse STATIK," a tentative name for an improvement technique that starts at step 5 (with the current designs of kanban systems) and backtracks through the preceding steps until the need for a change is identified. This change is then worked through to steps 5 and 6 in the conventional, forward direction.[60]

Further Reading

Achouiantz, Christophe. 2012. *Assessing the Depth of a Kanban Implementation* http://www.slideshare.net/ChrisAch/depth-of-a-kanban-implementation

---

60. I recently described "Reverse STATIK" in a post on the djaa blog "Reinvigorating an existing Kanban implementation with STATIK." http://www.djaa.com/reinvigorating-existing-kanban-implementation-statik

Anderson, David J. 2010. *Kanban: Successful Evolutionary Change for Your Technology Business.* Sequim, WA: Blue Hole Press.

Anderson, David J. 2012. *How Deep is Your Kanban?* http://www.djaa. com/sites/ltdwip/DepthOfKanban.pdf

Anderson, David J. 2012. *Lessons in Agile Management: On the Road to Kanban.* Sequim, WA: Blue Hole Press.

Benson, Jim and Tonianne DeMaria Barry. 2011. *Personal Kanban.* Seattle: Modus Cooperandi.

Brodzinski, Pawel, blog posts tagged "project portfolio" http://brodz-inski.com/tag/project-portfolio

Burrows, Mike, blog posts tagged "portfolio" http://positiveincline.com/index.php/tag/portfolio-2/

Carroll, Ian, blog posts tagged "portfolio management" http://iancarroll.com/category/portfolio-management/

Forss, Håkan. 2012–13. *Kanban Kata.* http://hakanforss.wordpress.com/tag/kanban-kata/

Hammarberg, Marcus and Joakim Sunden. 2014. *Kanban in Action.* Shelter Island, NY: Manning Publications.

Leopold, Klaus. 2014. *Kanban Flight Levels.* www.klausleopold.com/kanban-flight-levels

❖ Chapter 17 ❖

Smaller Models

This chapter describes a number of models that support the concepts covered in Part I and are helpful background for Part III:

◆ *Little's law*, a beautifully simple formula with a nice visual interpretation (and an excuse to revisit cumulative flow diagrams)

◆ The *Satir change model*, the late Virginia Satir's powerful description of the change process

◆ Two coaching models: the very useful thinking tool *GROW*, and Toyota's *A3* (first mentioned in Chapter 14)

◆ Jeff Anderson's *Lean Change Canvas* via a digression into the *Pyramid Principle*

◆ Various models of facilitation, including games

◆ Two models of **leadership** and **collaboration**, T-*shaped leadership* and *triads*

Two That Got Away

Little's Law

Some are no doubt surprised that this is the first real introduction of Little's law. Why, in Chapter 2 (**balance**), did I mention it only in a footnote? And not at all in Chapter 5 (**flow**)?

It is not that I am afraid of mathematics (I have a degree in it). Nor do I worry that it is difficult to explain (it isn't). But I am skeptical of anyone claiming to have proven from scientific principles that their methods must undoubtedly be the best, and I will not make that same mistake. I'd rather let the mathematics help you consolidate what you already know. Conventionally, Little's law is written as:

$$L = \lambda W$$

Where:

+ *L* is the long-term average number of work items (called "customers" in *queuing theory*) in a *stable system* (a strict definition of which isn't necessary here).
+ λ is the long-term average rate of arrival of work items.
+ *W* is the average time that work items spend in the system.

We tend to replace the symbols with words, using the overhead bar notation to denote averages:

$$\overline{(\text{WIP})} = \overline{(\text{Delivery Rate})} \times \overline{(\text{Lead Time})}$$

In swapping "Delivery Rate" for "Arrival Rate," we are reminded to be careful to account for work that disappears into black holes or materializes out of nowhere (knowledge work can be like that). Note also that sometimes we omit the bar over the WIP, reasoning that in a WIP-limited system, the WIP is effectively constant (this maneuver makes my inner mathematician a little uncomfortable, but it does make a useful point).

Little's law is the one piece of queuing theory I wish every manager knew, but if you remember only one thing about it, it is that these three quantities are intimately related. There is no escaping the math: You absolutely cannot change WIP, lead time, or delivery rate without there being some corresponding change in at least one of the other two. This is very good news, and it explains how a lot of system interventions actually work.

Let me show the formula as it is typically arranged in a Kanban presentation:

$$\overline{(\text{Lead Time})} = \frac{\overline{(\text{WIP})}}{(\text{Delivery Rate})}$$

Suppose we want to reduce lead times. Little's law helps us organize three possible lines of attack:

1. We can take the direct route, removing sources of delay from the timeline. This might be done Lean-style, removing non-value-adding activities, or Agile-style, replacing handoffs with something more collaborative. Little's law tells us that if we're successful, we

must see WIP decrease or the delivery rate increase. From experience, we're likely to see both.

2. We can take the indirect route of limiting work-in-progress. So long as any accompanying loss of delivery rate isn't too great (aggressive WIP reductions do carry that risk, but it's also quite possible that we will receive a double benefit instead), Little's law tells us that average lead times must come down.

3. We can take what might be called the Theory of Constraints route, increasing the delivery rate by adding people, adding other capabilities, or removing things like rework and failure demand that consume capacity unproductively. Little's law reminds us that we must implement these in such a way that the extra capacity isn't immediately absorbed by an equivalent increase in WIP (which is easier said than done, sometimes).

Let's rearrange the formula one last time, making the delivery rate its focus:

$$\overline{(\text{Delivery Rate})} = \frac{\overline{(\text{WIP})}}{\overline{(\text{Lead Time})}}$$

Because the delivery rate is by definition a quantity over time, just like the right-hand side of this arrangement, Little's law has a nice geometric interpretation. Let's overlay some triangles on the cumulative flow diagram (CFD) that we first saw in Chapter 1 (see Figure 17.1).

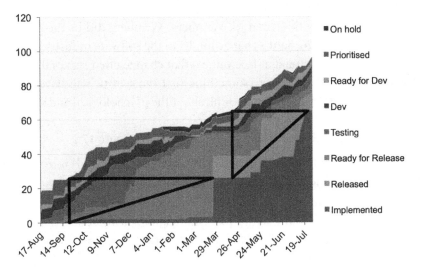

Figure 17.1   A geometric interpretation of Little's law

In the interest of clarity, I have removed the "Proposed" state from this version of the diagram, visualizing the process from "Prioritised" (*sic*) and "Implemented." Don't worry, though, I'm not cheating: Little's law can be applied to any part of the process so long as we can sensibly define arrival and delivery without losing track of items in between.

The vertical heights of these triangles are point-in-time measures of WIP. Their slopes are delivery rates, and their horizontal widths give an indication of the time it took to convert that amount of WIP into delivered items. These aren't the long-term averages of Little's law, and the CFD says nothing about which items got delivered when, but they are useful visual guides nevertheless.

You can see that the right-hand triangle has a steeper slope and shorter horizontal width than the left-hand triangle. Faster and quicker: This is real progress! I would have liked to see less WIP here, too, but you can see that a large proportion of this was in the "Ready for Release" and "Implemented" columns, and the WIP in preceding stages was under control. As described in Chapter 4 (**customer focus**), we knew we had issues in those later stages and were taking steps to deal with the underlying causes.

The Satir Change Model

Chapter 7 introduced the J curve. One particular analysis of the J curve effect is due to Virgina Satir, a distinguished family therapist and author who collaborated with Gerald M. Weinberg. Weinberg did us the good service of introducing Satir's change model to the software industry.

Instead of just taking it at face value—that change often makes things worse before they get better (something that most of us will have observed), Satir's model helps us to think about the psychological and social impact of change.

Satir describes change in five stages, as shown in Figure 17.2.

1. The *late status quo*, the period before the (unanticipated) introduction of the *foreign element*.

2. A period of *resistance*; the foreign element has been introduced but we try to cling to the status quo. Our heightened awareness may actually improve performance for a time, though unsustainably.

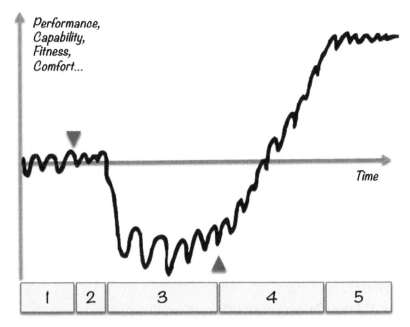

Figure 17.2   The Satir change model

3. We enter *chaos*; we are not dealing adequately with the foreign element and the status quo is no longer tenable.
4. *Integration:* The penny drops, the pieces fall into place, a *transforming idea* allows the foreign element to be taken on board.
5. We have a *new status quo* once this process of integration is complete.

Of course, not all foreign elements are benign, and the new status quo isn't necessarily better than the old one. In this regard, the idea of stages in Satir's descriptive model is more realistic and more helpful than the curve.

Thinking Tools and Coaching Models

These models support the kind of mentor conversations described in Chapter 8 (**agreement**). They are also very useful as thinking tools—ways to organize, test, and refine our thoughts before we share them.

GROW

I first became aware of the GROW model through John Whitmore's excellent book *Coaching for Performance*. GROW is a simple structure that can guide a coaching conversation, spoken or written:

- ◆ Goal
  - ◆ What would you like to have happen?[61]
  - ◆ What will "done" look like?
  - ◆ (Digging into a stated goal) What are the drivers for that?
- ◆ Reality
  - ◆ What is currently happening? How do you interpret the available data?
  - ◆ What is the impact? How could you quantify that?
  - ◆ Can you identify some likely root causes?
- ◆ Options
  - ◆ What could you do? What should you do?
  - ◆ How have similar things been achieved in the past?
  - ◆ Who might help?
- ◆ Will (or Way Forward)
  - ◆ What will you do? What comes first? Why?
  - ◆ Who will be impacted? Other obstacles?
  - ◆ How does your plan address your root causes, meet its goals?

I now look for this structure in other models, including the next one, *A3*.

A3

As mentioned in Chapter 14, Toyota structures its change-related mentor/mentee conversations around an A3, a proposal constrained to fit on a single sheet of A3 paper.[62] There is no A3 template as such, but proposals typically include the following elements:

- ◆ Some context, perhaps a visualization of the current process, perhaps some qualitative or quantitative analysis of the *current condition*
- ◆ Some statement of how things should be—the *target condition*

---

61. Thank you, Bob Marshall, for that one.
62. A3 paper measures 11.7 by 16.5 inches.

- A list of some possible *countermeasures*, ideas that could mitigate things we don't like about the current condition and take the system to (or at least toward) the target condition
- A plan that outlines how the chosen countermeasure (or countermeasures) will be implemented

There is a strong correspondence between A3 and GROW, though the G and the R are reversed here. Like GROW, it's the focus for a conversation; one should expect that an A3 will undergo thorough examination and significant change as a result of the mentoring process.

Using "countermeasure" instead of "solution" is significant; it is considered very important that multiple countermeasures are considered; failing to do so may indicate that the mentee might be fixated on a particular solution at the possible expense of actually solving the problem. It helps also to guard against a specific form of lazy thinking, expressing the problem in terms of the absence of a favored solution.

Digression: Learn to Review Like a Pro!

I have read a lot of project proposals in my time, some no bigger than an A3 (a side or two of A4, a handful of PowerPoint slides perhaps), some much bigger. Regardless of size, early versions of most of them have been pretty awful.

What makes this so depressing is that reviewing these is a teachable skill. It is so teachable in fact that the excuse, "No one can effectively review their own work" is actually very lame.

If you regularly read or write proposals and other structured documents, make sure you read Barbara Minto's *The Pyramid Principle*. You will never look at another list of bullet points in quite the same way again! I can't look at any structured writing without instinctively invoking the MECE checks (pronounced "me see," and short for Mutually Exclusive, Collectively Exhaustive), and checking that it is organized and summarized for maximum impact.

In proposals, I look for:

- Some expression of context, a description of what is to be changed
- Why this proposal is important—goals or objectives
- Who we are doing this for, properly identifying customers—all of them, and not just their proxies (product owners, project managers, and the like)

◆ Statements of the customer needs that will be met

◆ A multiplicity of options (the guard against fixation and lazy thinking)

◆ An outline of the chosen way forward

The trick is first to review these elements on their own (and the more structured the presentation, the more that Minto's tools can be applied), and then to reconcile between elements. Insist that:

◆ There is some appreciation of context.

◆ The needs are real, from the various customer perspectives.

◆ The options will meaningfully meet needs in ways likely to be acceptable to customers.

◆ The chosen option (or set of options) meets overall objectives.

◆ The plan will deliver what is needed and is capable of being executed.

If proposals can be kept short—and Toyota seems to prove that they can—these checks can be done in minutes.

The Lean Change Canvas

Jeff Anderson has developed a change method centered on a *Lean Change Canvas* (Figure 17.3), a visualization that contains all the elements I expect to see in a proposal, and more. Instead of writing a document or drawing

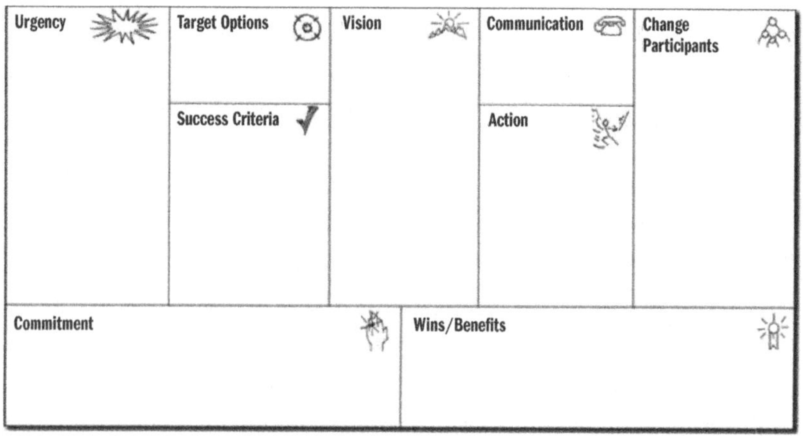

Figure 17.3   A Lean Change Canvas (courtesy of agileconsulting.blogspot.com)

up an A3, a canvas can be used to efficiently organize contributions captured in sticky note form.

Actions can then be copied to an auxiliary kanban board for execution. The method is described in Jeff's book *The Lean Change Method.*

Group Facilitation and Games

Kaner's Facilitation Model

In *Facilitator's Guide to Participatory Decision-Making,* Samuel Kaner describes the role of the facilitator with the model shown in Figure 17.4.

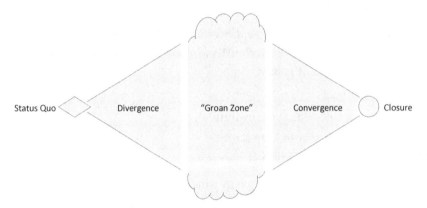

Status Quo   Divergence   "Groan Zone"   Convergence   Closure

Figure 17.4   The Kaner model of group facilitation

In this model, the facilitator guides a group of people through three stages:

1. *Divergence* ensures the plentiful supply of raw material—generating ideas, identifying problems, and so on.

2. Then comes the uncomfortable part, the "*groan zone,*" in which the raw material is sifted and analyzed. After the energy and creativity of the divergence phase, the discomfort of this phase stems from the lack of a clear way forward.

3. Finally comes *convergence,* in which thoughts become increasingly well organized and outcomes can be agreed upon.

The skilled facilitator creates the safety that encourages a diversity of contributions in divergence, supports the group through its discomfort (carefully preventing the *premature convergence* that prevents significant progress), and uses appropriate tools to shape the conversation toward valuable outcomes.

Based on this model, in *Collaboration Explained, Facilitation Skills for Software Project Leaders,* Jean Tabaka describes a number of meeting formats and collaborative tools applicable to development teams. It's a great guide to the dynamics of meetings and other collaborative encounters, full of advice to those who may find themselves in a facilitating role.

Serious Games

In her groundbreaking book *Reality is Broken: Why Games Make Us Better and How They Can Change the World,* Jane McGonigal describes four essential properties of games.

- ◆ A clear goal
- ◆ A set of rules (known constraints)
- ◆ Feedback (a way to keep score)
- ◆ Optional participation

What is striking is the degree to which these properties also describe (or would describe, if only things were better) a lot of what happens inside organizations. In *The Culture Game,* Daniel Mezick suggests that organizations become easier to change when you learn to look at what they do in this way. Mezick uses this model both to explain Agile methods and practices and to describe a social process of organizational adoption he calls *Tribal Learning.*

The generic term *serious games* refers to the deliberate use of games (many of which meet McGonigal's criteria) in order to achieve some serious purpose. The Agile community has adopted a number of them and has generated many more (there is a feast of them cataloged at tastycupcakes.org).

A simple and very useful game is *dot-voting,* which comes up again in Part III (I use it a lot). This brings up to date the ancient idea of *multi-voting,* which is simply a voting process in which participants each have a predetermined number of votes that they can cast against items—features, improvements, discussion topics, and so on—that need to be ranked. In the modern version, sticky notes represent items, and votes are cast by marking them with dots.

In *Innovation Games*, my friend Luke Hohmann combines *serious games* with a facilitation model—*Ideas Into Action™*—that builds on the Kaner model. Luke describes a number of games that work well on paper and translate very effectively to the Internet. Some games turn out to be significantly more playable with electronic support; I prefer the online version of the portfolio prioritization game *Buy a Feature*, for example. For other games, differences in playability are marginal, but facilitators benefit from the ability to analyze chat logs, generate reports on the exercise's findings, and so on.

With Luke's company Conteneo, Inc., we have created *Kanban Knowsy*, shown in Figure 17.5. In Knowsy games, participants guess each other's secret "like lists," a ranked top five chosen from a longer menu of choices. In the Kanban version, team members guess each other's rankings of the Kanban Method's principles, practices, and (most popularly) values. In doing so, they discover what is important to the team collectively, how well team members know each other, and how well they are aligned.

A game that works well both on paper and online is *Speedboat* (see Figure 17.6). In Part III I recommend this as a way for groups to visualize impediments. Not only can they be ranked without the need for a voting step (the vertical depth of the "anchors" is used to indicate their seriousness), it is possible to organize them horizontally, too, typically to represent the different stages of a workflow.

Figure 17.5   Kanban Knowsy

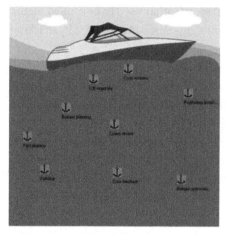

Figure 17.6   Speedboat (online)

Models of Collaborative Leadership: Triads and T-Shapes

The *Triad* is a very simple model of **collaboration** and collaborative **leadership** that has been practiced deliberately in a surprising variety of places. Thanks to *Tribal Leadership: Leveraging Natural Groups to Build a Thriving Organization*, the book by Dave Logan, John King, and Halee Fischer-Wright, we understand its applicability to corporate and community life. Triads appear in some churches in the form of *prayer triplets* (my wife, Sharon, has been a member of several of these); the model was even practiced by the KGB!

A triad connects three people, united by some common purpose. Sometimes it is the result of one person introducing two previously unconnected people; sometimes they are formed to perform some specific task. Effective triads obey two rules:

1. Each member takes some responsibility for the relationship between the other two members, providing strength.

2. Growth comes not by turning triads into quads, but by forming additional triads involving one or two members of existing triads, thereby creating networks.

I'm the kind of person who approaches a "networking event" with dread, and the triad model is just about the only form of networking that works for me. I have learned to make a point of introducing people whom I know to share some common interest. That's rewarding in itself, but often I reap double or triple the benefit in the form of fruitful collaboration and new introductions.

Triads express collaborative leadership when they are used deliberately to share knowledge, to create opportunities, and to form bridges between different parts of the organization. I have encouraged graduate recruits to form long-lasting triads and to help one another to grow their networks from them, and I have used them short-term to solve specific problems.

Morten Hansen describes *T-shaped management*, which is somewhat analogous to the *T-shaped people* I alluded to in this book's preface. His T-shaped managers encourage collaboration in two quite distinct ways:

1. Much in the manner described in Chapter 3, close collaboration inside their part of the organization
2. Addressing the downsides of collaboration described at the close of Chapter 3, "disciplined" collaboration across the wider organization

The key to Hansen's model is that this second kind of collaboration is required to be purposeful and effective; it is not about networking for its own sake, and it is expected to deliver results in healthy proportion to the effort expended. Ill-disciplined collaboration may be worse than no collaboration at all.

Both models are entirely compatible with Kanban's *at every level* kind of leadership. Triads don't need to respect organizational boundaries at all, and those T shapes can emanate from anywhere. We can all do it.

Further Reading

Anderson, Jeff. 2013. *The Lean Change Method: Managing Agile Transformation Using Kanban, Kotter, and Lean Startup Thinking.* https://leanpub.com/leanchangemethod

Hansen, Morton. 2009. *Collaboration: How Leaders Avoid the Traps, Build Common Ground, and Reap Big Results.* Boston: Harvard Business School Press.

Hohmann, Luke. 2006. *Innovation Games: Creating Breakthrough Products Through Collaborative Play.* Upper Saddle River, NJ: Addison-Wesley.

Kaner, Sam. 2007. *Facilitator's Guide to Participatory Decision-Making,* 2nd ed. San Francisco: Jossey-Bass.

Logan, Dave, John King, and Halee Fischer-Wright. 2008. *Tribal Leadership: Leveraging Natural Groups to Build a Thriving Organization.* New York: HarperBusiness.

Mezick, Daniel. 2012. *The Culture Game: Tools for the Agile Manager.* Guilford, CT: FreeStanding Press.

Shook, John. 2008. *Managing to Learn: Using The A3 Management Process to Solve Problems, Gain Agreement, Mentor, and Lead.* Cambridge, MA: Lean Enterprise Institute.

Sobek, Durwark K. II and Art Smalley. 2008. *Understanding A3 Thinking: A Critical Component of Toyota's PDCA Management System.* New York: Productivity Press.

Tabaka, Jean. 2006. *Collaboration Explained, Facilitation Skills for Software Project Leaders.* Upper Saddle River, NJ: Addison-Wesley.

TastyCupcakes.org: Fuel for Invention and Learning. http://tastycupcakes.org

Whitmore, John. 2010. *Coaching for Performance: GROWing Human Potential and Purpose: The Principles and Practice of Coaching and Leadership,* 4th ed. London: Nicholas Brealey.

❖ PART III ❖

Implementation

Part III is modeled on the *Systems Thinking Approach to Introducing Kanban*, or *STATIK*. David Anderson devoted a double-length session at the 2012 Lean Software & Systems Conference in Boston to this approach,[63] and it forms the backbone of most foundation-level Kanban training.

By chapter:

18 Understand sources of dissatisfaction

♦ Capture internal and external frustrations from multiple perspectives

♦ Identify sources of variability

19 Analyze demand and capability

♦ Identify work item types, patterns of demand

♦ Describe and try to quantify how (and how well) demand is met

20 Model workflow

♦ Visualize and review the knowledge discovery process

♦ Discover classes of service

21 Within the system's capability, match service offerings to customer expectations

♦ Design kanban systems

♦ Visualization, WIP limits, and policies

22 Roll out

♦ Planning and shaping

♦ Pulling change through the system

---

63. A video of David's talk is available at http://vimeo.com/46272041

It is rather misleading to present this process as though it happens in strict sequence. STATIK tends to be highly iterative and exploratory, and it needs to be. That way, we don't just help smooth the introduction of kanban systems, we model and catalyze the evolutionary change of the Kanban Method.

Because it's iterative, we can use STATIK in at least two ways:

1. To get a new Kanban implementation underway

2. To review and refresh an implementation that is already established

STATIK helps to connect (or reconnect) a Kanban implementation to the needs of the organization. Alluding to the *agendas for change* in Chapter 10, I refer to this as *shaping the agenda*, and it's one of the most satisfying aspects of my work inside organizations. Often this corrects a previous oversight—it is all too easy to copy the practice of using a kanban board while failing to grasp its broader significance.

Whether or not you choose to hire an expert practitioner from outside your organization, do make this a group activity rather than a solo effort. STATIK requires that you capture multiple perspectives, and it is far better if these are represented in the room rather than merely guessed at. To attempt it on your own would be an opportunity wasted, potentially even damaging. Without **collaboration**, how else will you build **understanding** and foster **agreement**? Where's the **respect**?

❖ CHAPTER 18 ❖

Understand Sources of Dissatisfaction

Any kind of deliberate change needs two key pieces of context:

1. Its scope—some boundary around *what we do now*, within which the change will be focused—the "what" of the change

2. Its objective—an expression of what we hope to achieve from the change, relative to how things currently are—the "why" of the change

In the beginning, it is unlikely that either of these will be known in any great detail. Don't let that worry you unduly—it's much better to start by exploring the problem space than to try to nail down solutions prematurely. And let's be realistic about this: Scope and objectives are often determined more by what people feel is organizationally possible than by what is necessary. As change agents, we can find this very frustrating, but it's okay: Let's start with what's possible; the impossible we can do later!

We start with sources of dissatisfaction because they lead very quickly to something much more positive: a set of things that people might want to achieve. Additionally, when we take the trouble to explore properly why these dissatisfactions are a problem to people, important issues like scope and sponsorship tend to become much clearer.

This is not the time for isolationism. You will need to talk to people! Be prepared to go out and meet them or to bring them in; make them part of the conversation.

Two Perspectives

Given even just a rough idea of scope, we can easily identify two quite different perspectives:

1. The perspective of those working in the system—their first-hand understanding of the system itself, their impressions of how it is perceived externally

2. The perspective of those outside of the system (customers, higher-level management, providers of related services)—how it helps to meet their broader needs (and what those might be), their impressions of how their immediate needs are serviced

Because we are getting people to look both inward and outward, it is no disaster if we later decide that we got the scope boundary wrong. Wherever we place that boundary, we learn a lot a by reconciling the two perspectives and accounting for any important differences. Opinions on the placement and nature of the boundary may be revealing, too.

Two Questions

Much of what we are trying to achieve in this chapter can be boiled down to two key questions:

1. From your personal perspective and from what you perceive from others inside and outside, **what are the main sources of dissatisfaction with the system?** In other words, what needs (and whose) aren't met?

2. **What sources of variability and unpredictability would you highlight?** In other words, what frustrates you and the broader system as you try to deliver things of value with quality and timeliness?

Both questions are about needs and the system's current capability to satisfy them. Asked in terms like these, we stand a very good chance of establishing some drivers for change, and making it personal.

Notice how the second question, which is aimed mainly (but not exclusively) at internal participants, relates in an obvious way to the value of **flow**. You may find it helpful to pause and think about what other values underlie both questions, perhaps even to make them explicit. To me,

understanding, **respect**, and **customer focus** seem very close to the surface; in your particular context you may see others.

Formats

I have seen all of these work well, sometimes in combination:

◆ Going out and talking to people individually, circling back around as a bigger picture emerges.

◆ Gathering people together into one place for a workshop, asking those questions directly and exploring them informally.

◆ In a workshop or online setting, using facilitated exercises or games that help individuals and groups to generate and organize a large quantity and diversity of responses.

Each has pros and cons regarding the amount of information generated, the quality of exploration possible, and the amount of shared understanding generated. Personal and organizational taste will have a bearing, too. Used in countless group sessions, my go-to technique works like this:

◆ **One sticky note per issue**: Individual participants write their dissatisfactions down, one short phrase in bold marker pen per sticky note, each one identifying (usually anonymously) a single dissatisfaction. Variant: Do this in small discussion groups.

◆ **Gather and cluster**: The sticky notes are brought together and organized, either as they're added to the wall, or subsequently. Duplicates get stacked; other closely-related items are bunched together.

◆ **Name the clusters**: Choose a name for each cluster that fairly represents its content. When a good name can't be found, we reconsider the clustering and try again. These names are displayed on sticky notes of a contrasting color or size.

◆ **Dot-voting** (optional): Everyone gets a fixed number of votes—typically 3 to 5—which they cast by marking a dot on a sticky note. By agreement, votes are applied either to the clusters as a whole or to individual issues, and multiple votes by the same person on the same item are usually allowed. Items are then ranked by the number of votes received.

♦ The *Speedboat* game mentioned in Chapter 17 is a good alternative. Instead of ranking dissatisfactions by the dot-voting process, participants in groups of five to eight people must
collaborate on gauging their relative impact as visualized by the
vertical placement of the "anchors." This may trigger conversations that the group may wish to summarize when they report
their findings. Some horizontal organization is also possible,
either predefined or decided by the group.

You might need to use different techniques in combination. Bear in
mind that internal workshops organized for the benefit of a single con-
stituency (training classes typically fall into this category) necessarily
rely heavily on perceptions that are at best secondhand and at worst wild
guesses or wishful thinking. These shouldn't be allowed to go unchecked.

Another option is to rely on the opinions of a few key people. But however perceptive and influential they may be, this wastes an important opportunity for creating shared understanding. I recommend against this
approach unless its purpose is just to set the scene for a bigger exercise.

Do, however, make good use of any supporting information that might
already exist. In large organizations especially, it's often possible to get hold
of things like climate surveys, recent work by external consultants, results
from team retrospectives, and so on. Whatever you find, make sure that it
includes the right range of perspectives; be wary of premature conclusions;
and be ready to drill down where needed.

Organize and Explore

A good facilitator knows not to allow these information-generating exercises to converge prematurely on too narrow a range of results. At some
point, though, you will need to organize what you have and establish a
sense of key themes and priorities. But express these carefully. Stick to
the language of dissatisfactions and frustrations; don't describe solutions.

Watch out for solutions dressed up as problems—"Our problem is that
we don't have *X*" (where *X* is a solution). As a facilitator, it's essential that
you get behind these to the dissatisfactions and frustrations that people
feel personally.

Two very important solutions often get dressed up as needs. They invite some knee-jerk responses that I really wouldn't recommend:

♦ "We need better-defined processes." (Don't say, "Yes, that's what we're doing with Kanban.")

♦ "We need better-defined roles." (Don't say, "I really wouldn't recommend starting with those," and please don't use Chapter 9 as justification!)

If you find yourself in danger of being this glib or dismissive, take it as a sign that the problem space needs to be explored more deeply.

Some genuinely felt dissatisfactions may be worth exploring further in case they hide other concerns. Some common examples are these:

♦ Concerns over timeliness may hide concerns over communication, coordination, or quality.

♦ Concerns over staffing levels, workload, prioritization, and overall system effectiveness tend to be somewhat interchangeable.

Again, take care to explore needs without assuming particular solutions. This exploration pays off handsomely if the "concern behind the concern" reveals a "need behind the need." Perhaps the system is failing to meet needs for the simple reason that they have never before been articulated clearly enough.

Follow Up

Finally: Share, invite feedback, refine. If you're doing this in a workshop, much or all of this should happen by deliberate design on the day. Other formats are likely to require much more in the way of follow-up activity, perhaps in several rounds.

What will you have achieved?

♦ Shared **understanding** and **agreement** on dissatisfactions, the beginnings of a case for change that a wide range of stakeholders will already have bought into

♦ Likely, either some kind of sponsorship, or at least a very good idea of where to seek it

And you'll have plenty of raw material; you'll be needing that.

❖ CHAPTER 19 ❖

Analyze Demand and Capability

The previous chapter was all about context and perceptions, mostly keeping implementation considerations off the table. This chapter is about gathering some specific qualitative and quantitative facts about the current process that will inform the design of kanban systems.

If you can do this collaboratively, you will further reinforce the work of the previous chapter. If that's not practical, you will need to share and review what you find after you have done the legwork.

This analysis needs to generate both qualitative and quantitative understanding. Qualitatively:

◆ Understanding the different types of work involved will help you to identify the different variations in workflow that need to be managed.

◆ Understanding the different sources of work will help you to prepare for, shape, and manage demand as it arrives.

◆ Understanding why the work is needed will help you to understand the types of risk involved so that they can be managed appropriately.

Quantitatively:

◆ Understanding the quantity of work involved will help you choose a manageable granularity to visualize and control it.

◆ Understanding the gap between actual process capability (measured in terms of delivery rates, lead times, predictability, etc.) and the expectations of customers and the wider organization will highlight what kinds of improvements are needed.

◆ Quantitative analysis of work recently delivered, currently in progress, and waiting to be worked on may tell us where improvements are most likely to be found.

We follow the familiar pattern, starting with what comes out of the process and working backward from there.

Know What You're Delivering, to Whom, and Why

What

When introducing Kanban inside organizations for the first time, sometimes I like to ask my audience to write down on sticky notes what they're currently working on, and we gather them on the wall. The variety of responses I get is enormous:

◆ "Projects" or "Programs," as in, "we deliver projects and programs"

◆ "Analysis" (or the name of another function or activity), meaning "that's what I and my closest coworkers do"

◆ Descriptions of specific low-level tasks that contribute toward specific deliverables, for example, "Create currency table" (or even just "Currency table"), things that don't add much customer value on their own

◆ The names of specific customer deliverables—products, features, and so on—for example, "GBP swap curve," "Month-end report," "Claims service"

Although I would tend to steer the conversation toward deliverables and away from activities and tasks, none of these are bad answers. They all tell us something about how work is organized and the corporate language by which it is described; and organizing work is what our kanban board will need to do. In fact, we might get to use these sticky notes later to bootstrap our board design (see Chapter 22).

There is no harm in getting an informal head start into quantitative investigation here. Having identified what gets delivered, we can find out roughly how big the items are, how long the delivery process typically takes, and at what rate (per unit of time) they are produced.

Here is how two teams to which I have belonged might have described their "what":

1. "We support a number of recently built applications and are working on two more. At team level we work mainly in features (sometimes bug fixes) of around a couple of days' development time in size, taking a few days overall. We make releases into production as often as we can, typically multiple times per week, sometimes more than one per day. At a higher level, we like to think in terms of business initiatives; we're thinking less and less in terms of projects."

2. "We work on a large, globally distributed system with a globally distributed team. Our deliverables are called just that—'Deliverables'—and they represent large features split by component and organized by 'Change Request' and 'Region.' We do full global roll-outs every six weeks or so, making smaller regional releases between times as needed."

To Whom

"*To whom*" can refer either to the downstream activity or function, or to the ultimate customer. Both are relevant, but the second is usually much more interesting—customers best provide the "why," to which we'll come in a moment.

Meanwhile, make sure no one confuses "to whom" with "under whose direction." Some people find it hard to get past the notion that they deliver to projects, to project managers, or to line managers; sometimes this confusion appears to be institutionalized.

Here's how those former teams might have described their "to whom":

1. "We do our own releases into production and we do our own support. Our immediate customers in the business use our systems to perform risk analysis on the energy markets, to generate trading alerts, and to deliver reports to the company's paying customers on a daily basis."

2. "Server-side, we provide installable applications to our regional support teams who do the deployment into production (we remain on hand just in case advice is needed). On the client side, desktop applications are updated on demand (this is largely transparent to most users). We have users in Sales, Trading, Operations,

Risk Control, and Financial Control in most of the countries in which we operate. Salespeople and traders have customers and counterparties all over the world; we are connected to many of them via electronic exchanges and other networks."

Often there are multiple "to whoms," in which case it's worth finding out whether the workflow involved differs at all by customer.

Why

We've learned a lot already, but the "what" and the "to whom" are really just the setup for the big question: "Why?"

Let's start this time with some possible answers from two teams I have previously managed:

1. "Our paying customers consume energy (electricity, gas, and oil) in such large quantities that it is worth their while to hedge their risks by buying some of it months in advance or by purchasing derivative contracts. Our tools help them trade the right amounts at the right time, so timely information is vital. We're working hard both to broaden our market coverage (energy markets are very fragmented) and to keep ahead of our competition in terms of sophistication and timeliness."

2. "Bond trading is all about risk: interest rate risk, credit rate risk, counterparty risk, and so on, investing in it and hedging it away. And the amounts of money involved are staggering! No participant can afford to be out of the market for very long, certainly not when key announcements or offerings are taking place. Our users need a fast and reliable platform capable of dealing with the ever-increasing variety and volume of business that a top-tier institution transacts."

It so happens that I have worked in two industries that are focused on risk very explicitly, and these two "why" statements say or strongly imply much about the qualities—timeliness, reliability, capacity, and so on—required in their respective products and processes.

For different teams in different industries, getting to a useful "why" might be harder work, but it's worth it, especially if a sense of purpose seems in any way to be lacking.

Armed with a sense of "why" at the highest level, we can look at a representative sample of individual deliverables. Why does the customer need them? What is the impact to them of faster or slower delivery? Is our process and product strategy enabling us to meet their needs effectively enough?

Quantitative Analysis

The level of quantitative sophistication needed here will depend on your situation; everything you have done up to now should give you a good sense of what type of analysis is likely to prove important.

Don't allow yourself to be put off by an apparent lack of data; neither skip this step nor kick off a long project of instrumentation and data collection. It shouldn't take long to get at least some rough estimates of many of these things:

- How often do we deliver?
- How many items go into each delivery?
- How long do deliveries typically take, end-to-end? How variable is that? How predictable?
- In a typical delivery, what does the age profile of its constituent items typically look like? You can measure ages relative both to when items were originally requested and to when they were first committed to.
- How many items are currently in progress, and what is their age profile?
- How many items are yet to be started, and what is their age profile? Do you regard them as your *backlog* (waiting to be processed) or something more flexible (ideas, options, etc., to be selected from)?
- Are things getting worse (longer durations, more WIP, less throughput, more defects, unhappier customers) or better? Can you quantify that?

If you can, visualize! Run charts, as shown in Figure 19.1, work very well for quantities that vary between items or samples, even when the number of samples taken each day tends to vary.

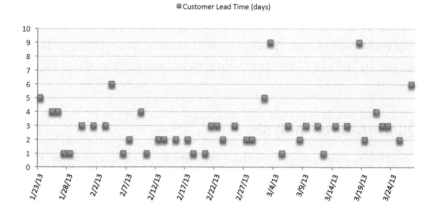

Figure 19.1   A run chart showing customer lead times

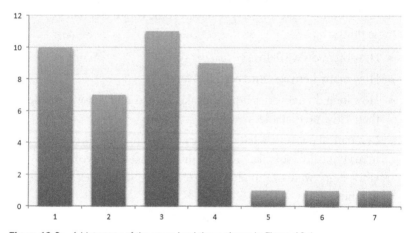

Figure 19.2   A histogram of the same lead times shown in Figure 19.1

Histograms summarize the data to reveal its distribution. Figure 19.2 is based on the same customer lead times as the run chart in Figure 19.1. It might be significant that this chart has two peaks (one at one day, the other at three); perhaps the data naturally separates into two distinct populations.

Pareto charts organize categorized data, such as work items by customer type or defects by root cause. They're useful when we want to prioritize our analysis. In Figure 19.3, customers A and B together account for more than 80% of the work; focusing on customers C, D, and E may not be very fruitful at this stage. You might be tempted to show this with a pie chart, but that wouldn't show the rankings nearly as clearly.

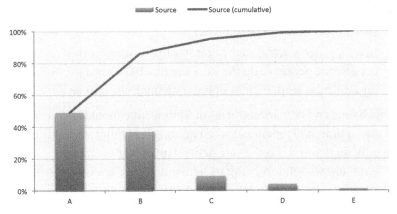

Figure 19.3   A Pareto chart

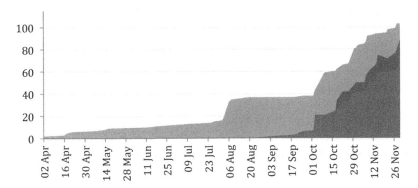

Figure 19.4   A CFD gathered from multiple sources

Even at this initial stage, a cumulative flow diagram (see Chapters 1 and 17) might not be out of the question. The early part of the CFD in Figure 19.4 was sourced from a SharePoint extract I managed to obtain. I was very glad to find it—I was prepared to trawl through several months' worth of meeting minutes if necessary. In the end it took me no more than a couple of hours' work to gather the data and visualize a 50-fold increase in throughput. Definitely worth the bother!

Flow Efficiency

One highly impactful metric is *flow efficiency*, which is the ratio of *touch time* to the overall lead time between defined start and finish points, expressed as a percentage. The touch time is the total amount of time a work item spends actively being worked on, as opposed to:

♦ In an explicitly inactive state, such as waiting in a queue or blocked (waiting for issues to be resolved)

♦ In a supposedly active state but not actually being worked on because the people to do the work are otherwise engaged (working on a different work item or doing something else entirely)

Touch time and the various forms of waiting time should add up to the overall lead time, and remember when quoting lead times to qualify them so that the start and end points are defined unambiguously.

It is not always valid to compare actual numbers between dissimilar processes (the metric is unhelpfully sensitive to the granularity of the work items, for example) but don't be surprised to find flow efficiencies that are significantly less than 100% and even at the single-digit end of the scale.

Low flow efficiencies tell you that there is a lot to be gained from removing waiting from the flow. Calculate one as early as you can so that you have a baseline for later comparison.

This definition of flow efficiency treats all touch time as equal. You could choose to distinguish between activities that are clearly about knowledge discovery and those that aren't.

How Work Arrives

Having looked at what comes out of the process and gathered some information about what happens inside, we're ready to look at how work arrives.

For each type of work:

♦ How does it arrive? Do you actively solicit it, wait for it to come, or something in between—conversation, exploration, or negotiation, say?

♦ At what rate does it arrive (so many per day, per week, or per month)?

♦ Does it arrive with regularity (e.g., through a scheduled meeting) or randomly (e.g., through calls to a helpdesk)?

♦ Does work arrive in large batches or bursts? Are there predictable seasonal variations?

♦ What are your key measures of success? How well do they align with customer outcomes?

My former teams might answer these questions like this:

1. "There isn't a regular pattern to work arrival. Our larger projects were started by agreement with the management team; the smaller stuff arrives through private conversations or via our team inbox. We aim to resolve support issues within 24 hours (and we perform very well by that measure); expectations on development work vary according to size and need."

2. "New work is managed regionally, typically through regular meetings held by regional managers or senior business analysts. Larger pieces of work are tracked on two weekly global calls (one for each of our main business lines) involving the management team and our counterparts in the business. Work is scheduled by release; work occasionally slips from one release to the next, but this is never without warning and discussion."

Does It All Add Up?

After this and the previous chapter's work, you will have a good feel for:

- What gets produced, in chunks of what size, and how often
- What gets asked for, how, when, and why
- How much work lies between, and for how long
- The dissatisfactions of those outside the process
- The frustrations of those inside

What do these tell you personally? Is there yet a collective sense of what needs to change? Or are there things that don't quite add up yet? What will you do with those?

Don't worry if a few ambiguities, disagreements, or gaps in knowledge remain—things will become much clearer once you're finished with the workshops and the analysis and you start to visualize work for real. The important thing is that the stage has been set.

CHAPTER 20

Model Workflow

This chapter looks at three different approaches to modeling the workflow that our kanban system is going to support:

1. Sketching it out
2. Top-down decomposition
3. Bottom-up organization

It's important to remember that the result that we're working toward is a working kanban system, not a static model. It's best not to get too attached to the products of this exercise—they will quickly lose their value once the system begins to evolve.

I'll keep this simple by assuming that there is just one main workflow involved. If you have more than one, you can take each workflow in turn or use one as the baseline by which the others are described.

Sketching It Out

We don't recommend any particular notation for this. It's up to you to decide whether you want to use a formal business process or *value stream mapping* (VSM) notation, stick to your in-house drawing conventions, or do something completely informal. Choose an approach that won't generate unnecessary resistance or cause you to invest more time on this step than is warranted.

You can afford to gloss over any detail that won't materially affect the decisions people make on what to work on and how. This consideration applies particularly strongly to functional boundaries and roles. Should their existence prevent people from making good choices? In an ideal world they wouldn't, and it's the job neither of our sketches nor of our eventual kanban systems to reinforce them.

Figure 20.1 describes, in a decidedly non-standard notation, the initial process of the XIT story (the case study of "From Worst to Best in Five Quarters," Chapter 4 of the blue book).

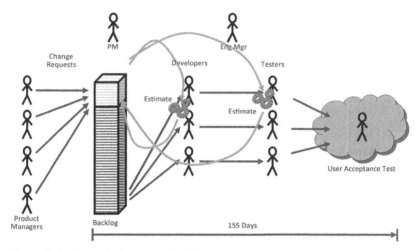

Figure 20.1   The original process of the XIT story

If you are familiar with the story, you know that the diagram attempts to show the disruption to developers and testers caused by the constant esti-mation of new items and the regular reprioritization of the huge backlog. Ten years later, however, I am tempted use a more linear representation of the journey taken by the work items, worrying much less about roles. For this purpose, Figure 20.2 seems to represent the process quite adequately.

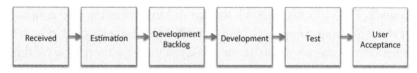

Figure 20.2   An alternative representation of the original XIT process

Are you left wondering what happens after "User Acceptance"? Good question!

Top-Down Decomposition

This sounds grand, but it amounts to writing down a very rough answer via a couple of guiding steps, then refining a little. Simply:

1. Identify your *commitment points* (one or two would be typical); for example, the points at which you first commit to start building or servicing something and later arrange to deliver, deploy, or otherwise complete it.

2. Give category names to the states or activities before, between, and after those commitment points. "Backlog," "Engineering," and "Implementation" might work in the XIT example.

3. Using as many levels as seem necessary, break down those category names in a way that reflects the reality on the ground. For example:

   - Backlog
     - Received
     - Estimation
     - Prioritized
   - Engineering
     - Development
     - Test
   - Implementation
     - User Acceptance
     - Deployment
   - Done

4. Identify the queues where work waits between active states and refine as necessary. These can often be found near handovers.

   In this example, "Received" and "Prioritized" are waiting states. It is very likely that work will also wait between "Development" and "Test" if different people perform these activities. A "Development Done" or "Test Ready" state would indicate that work tends to wait before or after this handover.

A highly informal variation of this approach is to start with "To Do," "Doing," and "Done" (perhaps this kind of setup is in place already), then ask

♦ *Are there different degrees of "To do"? Do we organize by activity or by priority?*

♦ *What happens in "Doing"?*

♦ *Are there different degrees of "Done"? How do we find out that it is "really done"?*

Bottom-Up Organization

Instead of digging into the distinction between activities and work item states, I glossed over it with the catch-all of "category." That was sneaky but deliberate: Keep in mind that our goal is to find an effective organization of work items.

That goal is the basis of this third strategy. Instead of modeling the workflow, why not just organize the work items that we actually have? This approach can work well, but it depends on having enough work items to represent a sufficient variety of states.

You might do this by exploring your existing tracking system, but a more visual, (and more social) way is to take the sticky notes from the previous chapter's work and arrange them horizontally according to how complete they are. This is given not by the amount of time they still re-quire, but by the state the work is in (or equivalently, the activity most responsible for making progress on that work item). Questions like these can be helpful:

♦ *What does this item need?*

♦ *What will this item need before it is more like that one?*

♦ *What will this item need before it can be considered complete?*

♦ *How did this item get to here?*

Now it is a matter of giving names to the groups of items in similar states of completion, then grouping or consolidating these categories until you have a configuration that works.

Incidentally, this technique can highlight some interesting cases:

Q: *What does this item need?*

A: *It needs a bug fix before testing can continue,* or

A: *It needs to come back from Team X*

Chapter 22 returns to these examples; it's important to pay particular attention to how we visualize rework and dependencies, two key sources of delay and frustration.

Review

Before we move on, here are some good ways to check and improve what you have so far:

- Quickly combine approaches to validate your model, such as:
  - Produce a sketch from your top-down or bottom-up model.
  - Make sure that actual work items map to your sketch or top-down model, then use the *"what does this item need?"* questions.
  - Consider whether it would be helpful to group, consolidate, or break down categories.
- Check that queues are adequately represented and that you know where your commitment points belong.
- Look to see where the dissatisfactions and frustrations discussed in Chapter 18 might originate.
- Identify the kinds of knowledge discovery associated with each active state.
- Seek to de-emphasize functional organization.
- Present it to other people.

None of this should take long. Remember, sketchy!

Chapter 21

Discover Classes of Service

Chapter 2 talks briefly about *classes of service*, describing them as categorizations associated with customer expectation and schedule sensitivity. To review, here are the descriptions of four categories that occur so often that we have given them standard names:

- *Expedite*: work items that are so urgent that we will drop other work in order to give them immediate attention.

- *Date-driven* (or *Fixed Date*): work items whose delay beyond a specific date will result in a significant penalty being incurred, disproportionate to any benefit in delivering early. Their schedule risks need to be managed actively.

- *Standard*: urgency-driven work, to be delivered in some customer-agreed order or sequenced according to a system policy. Suitable policies can be as simple as *first in, first out* (FIFO) or *let team members choose*, or based on an economic model such as *cost of delay* (see Chapter 15).

- *Intangible*: system improvements, maintenance upgrades, experiments in technology or the market, investments in people—work that's essential over the medium to long term but whose direct and immediate business value is hard to quantify.

To the system designer, classes of service are about embracing variety (as opposed to denying it or over-organizing for it) so that predictability can be improved. Their importance goes well beyond system internals, however; they're a visible part of the service offering and the means to bring customer expectations and system capability into alignment.

Those standard classes are really just to get you started. The trick is to identify qualitative differences that

1. Can be easily and uncontroversially recognized in advance
2. Require different handling, scheduling, and/or risk management internally
3. Will influence reasonable expectations externally

The concepts of classes of service and work item types (the "what" of Chapter 19) clearly overlap. The distinction becomes interesting when choice is involved, for example:

- Is this "support" work (with a formalized service expectation), or "small maintenance" work (joining a queue of variable length, perhaps to be prioritized as a later step)?
- Is it needed "by Friday" (to be risk-managed as a date-driven item), or "as soon as possible" (on a first-come-first-served basis, say)?

Without knowing the needs that lie behind a request, who can say? Some investigation and negotiation may be appropriate, but in many cases the customers are trusted to choose for themselves.

Discover, Check

Look at each work item type in turn. For each one, ask the following:

- Are all items treated as though they are equally urgent? If not, how are the more urgent ones identified? Is there any special language associated with them?
- Does the data on past performance suggest that there are multiple populations making up the whole? If so, how reliably can they be identified by factors easily recognized up front?
- What choices are offered to the customer? To what extent are work item types interchangeable?
- Are there *Service Level Agreements* (SLAs) in place already? Are they effective?

If the answers to these questions suggest that there may indeed be multiple classes of service, check that they will be valuable in practice:

- How will internal behavior be influenced by their presence? What policies would apply, and what would the impact be?
- How might customer behavior change if multiple classes of service were offered? How would customers benefit?

To double check, look at the current workload:

- What items are getting (or would deserve) special treatment?

Examples

These examples are typical of the range that might be encountered even within a single implementation:

- Support requests with a service expectation of a 24-hour turn-around
- Upgrades to external interfaces:
  - Released on specific dates, for reasons of business continuity
  - Treated as urgency-driven items, for reasons of business opportunity
  - Regarded as intangible items, for the sake of system health
- Major technology upgrades (e.g., a version upgrade of the underlying database technology or language platform):
  - Kept ticking over in the background as a series of intangible items and released when ready, well clear of any high-impact deadline
  - Run as a major project at the "last responsible moment" (inevitably pushing other projects out of the way)
- Requests identifiable in advance as likely to require clearance from Security, Operational Risk Control, or a similar veto-holding group

Toward a Healthy Mix of Work

This much should be obvious: You can't plan to have everyone occupied 100% of their time on fixed-date work, allow them to be interrupted by expedited items, and still expect predictability. Even without the interruptions, 100% utilization on work that must meet stringent deadlines is a recipe not for efficiency, but for extreme unpredictability and delay, not to mention pain at the human level. Systems need *slack* if they are to absorb variation rather than accumulate it, and people need it, too.

You might be wondering what a "safe" (or "humane") utilization level might be. 90%? 80%? I believe that this approaches the problem from the wrong direction. From direct experience as a manager, and from clients sampled by myself and David, fixed-date items need make up no more than 20% of the workload of most organizations; often, it can be significantly less. After allowing (say) 10–20% for intangible work, and a realistic provision for expedited and other unplanned work, the biggest category of all should be the high-value, urgency-driven work. This is exactly as you would hope it to be, assuming that you seek to maximize the flow of value to the business. Happily, predictability is achievable, too, thanks to the presence in the systems of work items that can safely be delayed when time-critical work needs to take precedence.

Classes of service and other categorizations enable some broad-brush prioritization decisions to be made independently of potentially contentious decisions about individual items. Making these explicit creates the opportunity both to align them to wider corporate priorities and to match them to customer needs. So long as there is the short-term flexibility to trade between categories, it should be possible to satisfy both constituencies.

For your workshops, questions like these help guide the overall mix of work in the right direction and prompt discussion about system design:

◆ What proportion of our portfolio is necessarily date-driven? Are we entering into date commitments unnecessarily, and if so, can we begin to bring this proportion down? Do we make reference to available capacity when scheduling? Are we able to respond when dates are under threat?

◆ How much capacity do we need to hold in reserve for expedited and other unplanned work? What steps can we take to reduce and/ or limit the amount we are servicing at any given time?

◆ Are we doing enough intangible work? What space do we give people to explore better ways of doing things?

◆ What capacity remains for standard, urgency-driven work? Do our systems reliably bring the right items to the front of the queue first?

If these questions make you uncomfortable, perhaps they should! Could it be that people are over-organized—"projectized" or siloed? If your organization lacks the ability to self-organize when and where it most needs it, it is paying a heavy price.

I have in the past described Kanban's ability to deal with variety as one of its "killer features." Arguably, the way STATIK encourages organizations to embrace variety is even more powerful. If only it were better known!

❖ CHAPTER 22 ❖

Design Kanban Systems

Y ou've got all the pieces lined up; now, for the fun part—board design, WIP limits, and the rest.

Scope, Work Item Granularity, Work Item States

These three parameters are best decided together. Make the scope of your first board too large, the granularity of its work items too small, or their states too short-lived, and you might end up with a board design that is too busy to be practical. Conversely, a board that is too narrow in scope and deals only in large items whose states change only very slowly isn't going to be very helpful.

These choices need to be compatible with the board's purpose and audience. A board designed to help manage the workload of individual engineers probably doesn't belong in the CIO's office, although some kind of project-level board might. At a more mundane level, your options are also constrained by the physical limitations of board size and location (these apply even to electronic systems).

Ideally, you want each member of a board's target audience to identify with at least a few work items at any given time, and for the board as a whole to evidence some meaningful movement between standup meetings. If these aims appear to be irreconcilable, you may need either multiple boards or a multi-level board design.

Sequential States

In Figure 22.1, I've taken an example from Chapter 20 and translated an outline of bullet points to a basic board design. It is very straightforward: Work items arrive in "Received" and move from left to right until they reach "Done." At some suitable time later they can be archived from the board.

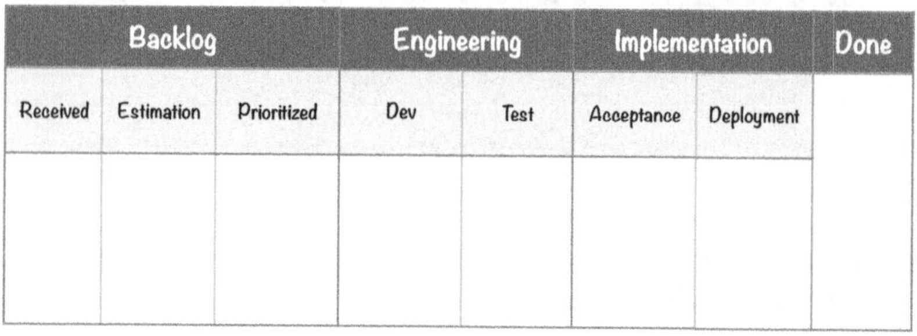

Backlog			Engineering		Implementation		Done
Received	Estimation	Prioritized	Dev	Test	Acceptance	Deployment	

Figure 22.1   Sequential states

The states have some high-level organization ("Backlog," "Engineering," "Implementation") but this need not have any material impact on the way the system is operated.

Parallel States

It isn't always possible to arrange states in a strict left-to-right sequence, particularly when some state changes happen in parallel with the main process of knowledge discovery. For example:

◆ Work may proceed optimistically, allowing approvals to be recorded after work has already started.

◆ Work may become blocked on a technical, quality, or business-related issue at any stage in the process; some degree of rework may be expected.

The most appropriate visual techniques for dealing with parallel states are often ticket-related:

◆ Recording approvals by marking tickets, perhaps incorporating a checkbox in the visual design of the ticket as a reminder

  ☐ Performance tested
  ☑ User approved

◆ Similarly, marking tickets to show that some optional ancillary activity will be needed, and again when that is complete

▤▤ UX Design

☑□ Database migration

As in Figure 22.2, prominently flagging blockers by overlaying the blocked ticket with another of a suitably alarming color (pink being the conventional choice for these "blocker stickies").

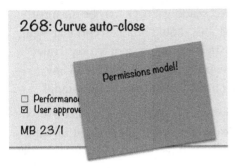

268: Curve auto-close

Permissions model!

□ Performance
☑ User approve

MB 23/1

Figure 22.2   Blocked!

Alternatively, instead of visualizing these state changes, handle them away from the board with policies such as column-level *definitions of ready* or *definitions of done*, making sure that there is agreement on how they will be observed.

Defects

The "blocker sticky" technique is very often used to indicate the presence of defects. I will admit to some strong personal preferences here:

◆ I really don't like to see work items advance through the process when they are known to be defective. This is storing up trouble for later!

◆ I get just as uncomfortable when I see work items that require rework being moved backward rather than getting blocked in place. Often this results in the true state of the work being misrepresented, sometimes dangerously.

But I will make exceptions:

- ◆ Defects that belong more to the existing product than to the work item in question might quite reasonably be raised as new work items, allowing existing work to continue unimpeded.

- ◆ Occasionally it makes sense economically to allow work items to proceed even when they are known to have issues. But be careful: Make this call too often and the quality of the overall product or service will quickly degrade.

- ◆ If it is clear that the work item should never have been considered fit to advance in the first place, sending it back may be the most appropriate course of action. This should be an exceptional event, however, one worthy of investigation. What went wrong? How does our process allow it to happen? Have we left behind any wreckage that needs clearing up? What didn't we understand?

Hybrid solutions may be appropriate. For example, if a defect is discovered in a feature that has been released but is still in its validation state or warranty period, its still-visible ticket can be flagged in place and a new work item raised for the rework. This accurately represents two important facts: We have a defective feature out there in production, and its fix is progressing through the system.

Dependencies

This covers two concepts: dependencies between work items, and work items that require attention from other services. Two main techniques apply:

1. Show the dependent work item as blocked, as in Figure 22.3. When, as is often the case, this situation is foreseeable rather then exceptional, use an appropriate color-coding, perhaps a color per service. Where the dependency is on another work item, identify it by name (and its ID number, if you use those).

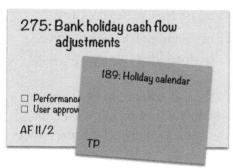

275: Bank holiday cash flow adjustments

189: Holiday calendar

☐ Performanc
☐ User approv

AF II/2
TP

Figure 22.3   One item dependent on another

2. Move the work item to a holding area, perhaps named after the service in question ("Infrastructure," say).

The first technique is usually the better choice for dependencies between work items. For dependencies on services, either technique will work; the second one makes the board design more complicated and may not always be feasible, but it does make it easy to see how many items are in that state, perhaps even to track some lower-level workflow.

Other Dimensions

The greater the number of work items to be visualized, the more useful it becomes to organize them by additional dimensions, such as:

- Work item type and/or class of service
- Source
- Some kind of less permanent category—initiative, project, epic, sprint, and so on
- The work item's "owner"

The two main choices for representing these dimensions visually are these:

1. At ticket level, using some combination of ticket color, annotation, or adornment. For example:
   - A default ticket color of yellow for "standard" (urgency-driven) work items
   - Amber tickets for date-driven work items, annotated with the due date
   - Red tickets for expedited work items (or no ticket at all, if they're short-lived enough)
   - An annotation denoting the sprint for which the work item was committed (I like to write the sprint number on small colored labels for this, using a different color per sprint so that they can be recognized from a distance.)
   - Marking tickets with the initials of their owners, or placing a movable *avatar* on the ticket

2. Adding lines to demarcate horizontal *swim lanes* across the board or part of it. For example:

   ◆ A permanent swim lane dedicated to expedited items

   ◆ Swim lanes dedicated to teams or people (I discourage this if it is likely to impede self-organization)

   ◆ Swim lanes that organize work items by project, initiative, and so on, as in Figure 22.4. The names of these swim lanes will be temporary.

Stream	Backlog	Engineering	Implementation	Done
A				
B				

Figure 22.4   Two parallel work streams

Multi-Level Designs and the Expand/Collapse Pattern

Consider the more complex design shown in Figure 22.5.

Let's break this down. The early part of the board (the left-hand side), shown in Figure 22.6, deals in large *epics*.

The later part of the board (toward the middle), as shown in Figure 22.7, allows up to three epics to be *expanded* into smaller work items (*features* or *user stories*).

This particular design is built on the assumption that features can be deployed and validated independently. Where this is not the case, we *collapse* them back into epics that can be delivered whole. A board design that supports this *expand/collapse* pattern is shown in Figure 22.8.

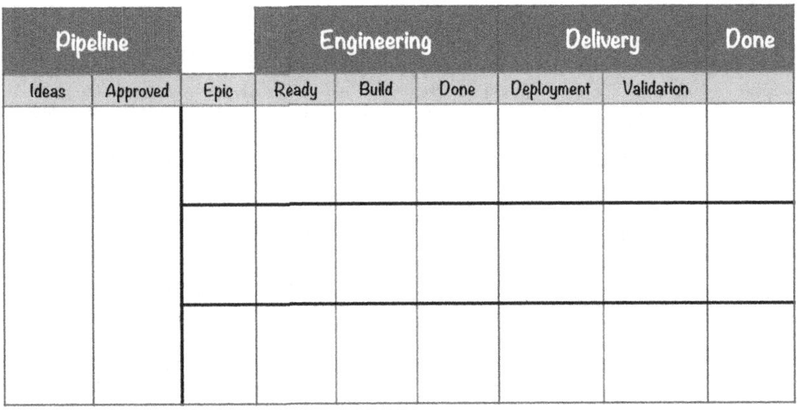

Pipeline			Engineering			Delivery		Done
Ideas	Approved	Epic	Ready	Build	Done	Deployment	Validation	

Figure 22.5   Epics and features

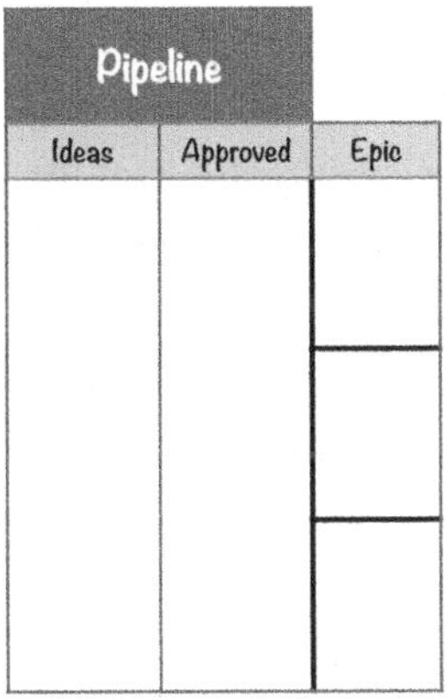

Pipeline		
Ideas	Approved	Epic

Figure 22.6   A pipeline of epics . . .

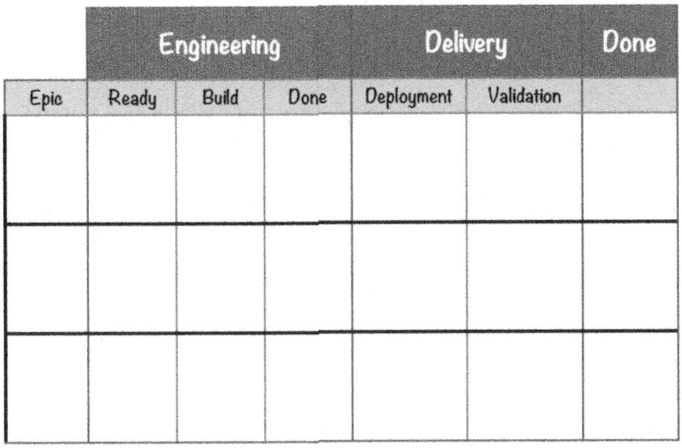

	Engineering			Delivery		Done
Epic	Ready	Build	Done	Deployment	Validation	

Figure 22.7   . . . Delivered feature-wise

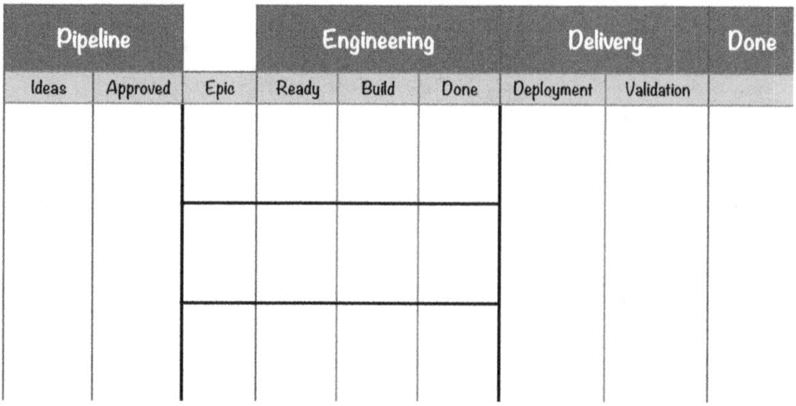

Pipeline			Engineering			Delivery		Done
Ideas	Approved	Epic	Ready	Build	Done	Deployment	Validation	

Figure 22.8   Expand/collapse

Limiting WIP

A true kanban system incorporates some mechanism for limiting the amount of work in progress. Once again, there are numerous ways to achieve this:

◆ Numeric WIP limits, each controlling the amount of WIP in a single column, a span of columns, or a horizontal swim lane.

◆ Physical constraints, such as the number of horizontal swim lanes available.

◆ Limits on the number of tokens (e.g., personal avatars) in circulation, and attaching them to tickets. The control offered by this mechanism is lessened if tickets are allowed to be in progress without such a token; in this case, it is limiting not the overall WIP, only the amount considered "active."

◆ Policies that control, for example, the amount of WIP per person or class of service.

◆ Some external mechanism that releases work into the system only when there is capacity. Scrum achieves this by limiting the amount of work committed to in a timeboxed sprint.

There is a lot to be said for using multiple techniques in combination, for example:

◆ Horizontal (e.g., class of service) and vertical (state) limits in combination

♦ Physical constraints (e.g., on epics) combined with vertical limits (e.g., on features in active development)

♦ Sprints combined with column-level limits, to curb multitasking and to encourage work to move from states of partial completion (e.g., "code complete") and on toward deployment

When multiple crosscutting mechanisms have been properly calibrated, no limit needs to feel overly constraining on its own, and yet the combined effect is to reduce WIP to levels that might previously have seemed inconceivable. Each limit addresses a particular symptom or behavior, kicks in when needed, and helps to smooth flow overall.

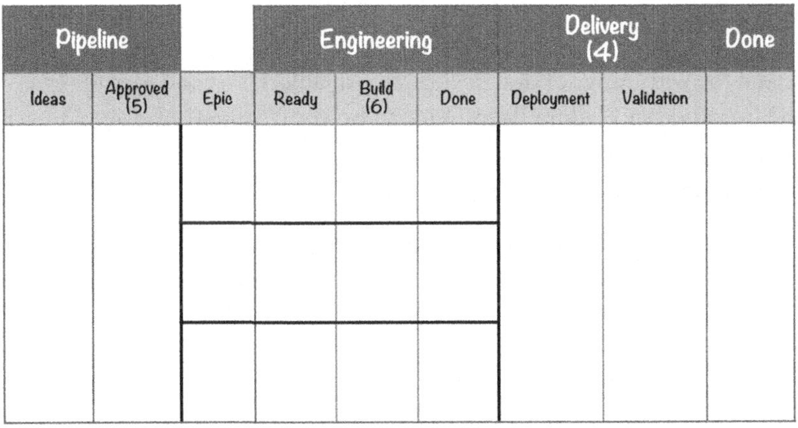

Pipeline			Engineering			Delivery (4)		Done
Ideas	Approved (5)	Epic	Ready	Build (6)	Done	Deployment	Validation	

Figure 22.9   Crosscutting WIP limits

To see how this might work in our expand/collapse example, consider Figure 22.9. Here we have a numeric limit of 5 epics in the single column "Approved," a physical constraint of 3 epics through "Engineering," and a numeric limit of 4 epics through "Delivery." This limits to 7 (i.e., 3 + 4) the number of epics actively in progress and to 12 (i.e., 5 + 3 + 4) the number of epics committed to and not yet complete. The limit of 6 features in "Build" spans the 3 active epics without dictating how they should be allocated.

Each of these limits has a different effect:

♦ The limit on "Approved" epics requires that some prioritization take place; if new items arrive with sufficiently high priority, items already waiting here may need to be pushed backward (not a bad thing under these circumstances).

♦ The constraint on epics in "Engineering" leads to a "one out, one in" kind of flow, even at this higher level.

♦ The limit in "Build" should encourage features and their respective epics to be built quickly.

♦ The limit on epics in "Delivery" should prevent them from hanging around for longer than necessary. However, a larger limit and/or a limit on "Deployment" may be needed if the whole process is likely to block due to problems of customer availability.

Review

Given the wide range of design choices available, it's a good idea to review the board design before deeming it to be your "initial" one (don't ever think of it as "final"!).

Begin with some basic technical checks:

♦ How does the board look when populated with work items? Does it organize them effectively? Is there enough room?

♦ How much movement will we see between standups? Not enough? Too much?

♦ Is it understandable by its intended users? Too complicated? Too simplistic? Ask them, or better still, involve them.

♦ Does it make unreasonable demands or assume changes of process, organization, or role that aren't yet agreed upon? (I don't say "never" to introducing radical system changes at this stage, but I don't usually recommend it.)

More reflectively, a good design addresses multiple needs of a broader nature:

♦ It brings **transparency** to what is happening and how it happens, helps better decisions to be made, and encourages self-organization and **collaboration**. Can you (collectively) picture those things with this design?

- It helps to bring **balance**—between demand and supply, across different categories of demand, and so on. Will it do that, even if only tentatively?

- It encourages both thought and action with respect to **customer focus** and **flow**. Refer to Chapters 4 and 5 and try using your board design as a thinking tool.

Lastly, and crucially:

- In what ways does this design begin to address the specific dissatisfactions and frustrations you captured at the very start of this process (Chapter 18)? Don't expect to fix all of them right away—it may be better not even to try—but will it bring their symptoms or causes closer to the surface than they are now?

These review considerations—and most especially the last one—apply not only when designing a board for the first time, but when evolving it, too. Better an ugly board whose changes are deliberately targeted than a beautiful one whose refinements exist for their own sake. Learn to see through the board to the system it represents; act on one for the effect it has on the other.

CHAPTER 23

Roll Out

I find it useful to think of Kanban implementation as a three-stage process:

1. **Planning the engagement**: preparation in terms of participants, venues, tools, supporting material, and so on

2. **Shaping the agenda**: approaching STATIK with the explicit aim of producing a compelling set of agreed upon priorities, goals, and actions

3. **Pulling change through the system**: maintaining momentum into the future, ensuring that progress will continue to be both visible and meaningful

This structure can be applied regardless of whether your aim is to build a stand-alone kanban system, to introduce the Kanban Method for the first time, or to reinvigorate fresh cycles of change. You can even use it retrospectively, helping you to think constructively about an implementation that needs a stronger connection with its host organization.

I hope to show that there is no contradiction between introducing Kanban impactfully and remaining true to its humane ethos. Toward the end of this chapter we review the role that the values can play in motivating an implementation.

Planning the Engagement

When executing any facilitated process, it really pays to prepare properly ahead of time. As facilitator, these are the kinds of questions you need to ask yourself when planning to use STATIK:

1. Understand sources of dissatisfaction (Chapter 18).

   ◆ Do you have at least a rough idea of the scope of the exercise? Does it have (or need) sponsorship?

   ◆ Who should participate? Who will represent the people who work inside the presumed system boundary? Who will represent the system's customers? Does the wider organization need representation also?

   ◆ What tools will you use to solicit and organize a good range of responses (see Chapter 17)?

2. Analyze demand and capability (Chapter 19).

   ◆ Are you ready to help capture the "what, to whom, and why"?

   ◆ Will you ask for data in advance (and perform some initial quantitative analysis yourself, perhaps), or will you wait to see what other participants want to do? What support can you provide?

   ◆ What do you already know about how work arrives? Will you have the people you will need to explore this properly?

3. Model workflow (Chapter 20).

   ◆ What is your preferred approach (sketch, top down, bottom up)?

   ◆ Are you armed with searching questions for reviewing the output?

4. Discover classes of service (Chapter 21).

   ◆ Will you get more traction approaching classes of service as an internal tool for organizing and scheduling work, or as a way to explore customer expectations? How will you bring the two aspects together?

   ◆ At which organizational levels will this exercise be able to influence the overall mix of work in the system?

5. Design kanban systems (Chapter 22).

   ◆ How will you introduce the concepts and share what has worked elsewhere? (This question answers itself when the workshop is also a training class.)

   ◆ How much time will you want to spend refining designs before allowing them to be tried in the field?

   ◆ Physical or electronic? What limitations (physical, geography, organization, privacy and security, or feature-wise) will need to be accommodated?

Lastly, don't neglect mundane things like the choice and availability of venues, participants' schedules, stationery, and equipment. These things matter even when you don't plan to hold a big workshop.

Shaping the Agenda: The Three P's

We've planned the details; now we need to step back and think about how the exercise is framed. Positioning, purpose, and priority give some high-level shape to the engagement that potential attendees can respond to.

Positioning

How you choose to engage with the organization and its people will depend both on context and on your own preferences. Again, it pays to put some thought into this. You need to think about how you position the Kanban Method with respect to the needs of the organization (to the extent that you are aware of them).

To get you started, here is a positioning based on some familiar elements:

◆ The Kanban Method has been described as the humane, *start with what you do now* approach to change.

◆ We will briefly explore its principles, practices, and values, thinking about how they apply in our situation.

◆ We will take a look at *what we do now*—warts and all—using what is known as the *Kanban Lens*. This encourages us to recognize that:

　◆ What we do—our flavor of *creative knowledge work*—is service oriented.

　◆ Service delivery involves workflow.

　◆ Workflow involves a series of knowledge-discovery activities.

◆ Through a series of exercises, we will do the following:

　◆ Map our knowledge-discovery workflow.

　◆ Pay attention to how and why work arrives.

　◆ Equip ourselves to track work as it flows across and between services.

◆ To ensure this exercise's success, we will take time to:

　◆ Agree on the scope and purpose of the system under review.

- ◆ Identify sources of dissatisfaction, which we will do from multiple perspectives.
- ◆ Prioritize actions that begin to address those dissatisfactions and better align the design and operation of the system to its purpose.

Purpose

Do you remember STATIK-0 from Chapter 16? Plan to incorporate its step 0:

0. Understand the purpose of the system.

It might seem that the obvious place to address this is before step 1 (Understand sources of dissatisfaction), but I recommend that you dwell on the "what, to whom, and why" of step 2 (Analyze demand and capability) and use that as the springboard for exploring how effectively the system serves its purpose.

This sets you up nicely for a discussion on *fitness for purpose* and *fitness criteria*. This should be much more than just a philosophical aside; identifying gaps and measures of success will help focus the subsequent design activities and give additional impetus to the rollout.

Priorities

In a similar vein, I have found it very powerful to prioritize the values and identify a top three or four around which a compelling call to action can be built. I do this through the rather unimaginatively named "Kanban values exercise," materials for which I have published under an open-source license at SlideShare.[64]

Participants receive a list of the nine values; they map them first to the Foundational Principles and then to the Core Practices. Experience has shown that this order works best, and it's much easier if the principles and practices are taken in reverse order.

I like to explain that there are no wrong answers—but there is a canonical answer! If you follow my recommended mapping and work

---

64. You can download a PDF from http://www.slideshare.net/asplake/kanban-values-exercise and I'm happy to provide the source PowerPoint file on request. Both are licensed under a Creative Commons Attribution-ShareAlike 4.0 International License.

bottom-to-top, you'll be looking for **leadership**, **respect**, **agreement**, and **understanding** for the principles, then **collaboration**, **transparency** (twice), **customer focus** and **flow** (both for CP4), **balance**, and (for a third time) **transparency** for the practices. Doing the principles first allows four values to be crossed off before proceeding to the trickier set.

We then use two rounds of dot voting, allowing three votes per person per round, to identify

- ◆ The values that—for whatever reason—resonate with the most participants at a personal level
- ◆ The values that should be emphasized in the implementation initially

Discuss the results of each round. Differences between the two can be striking; in your role as facilitator, take care not to jump prematurely to negative conclusions if this is the case (I have made that mistake).

Finally, take the top three or four values identified in the second voting round and try to construct some kind of narrative around them. The Kanban agendas can be useful guides here, but the results should be specific to the context and "owned" by the group.

Groups of between five and eight people work best. If you have more than eight people, you can consolidate multiple outputs in your debrief.

Between the two rounds of voting you can host a Kanban Knowsy group game (Chapter 17). Pitch this as a fun way to find out how well people have been listening to each other as well as to explore how well they are aligned. You can feed results of the second round into Kanban Knowsy's "Discover Play" for comparison with the three predefined agendas.

Pulling Change Through the System

You've done the planning and the framing; you've held your workshop or done the rounds; priorities for change have been agreed upon; teams are settling into their new kanban systems. How do you ensure that it doesn't stop there?

We are fortunate now to have a good number of well-documented Kanban implementations of significant size out of which some great advice has been abstracted—see, for example, *The Kanban Kick Start Field Guide* by Christophe Achouiantz and Johan Nordin, Yuval Yeret's *Pull-based*

Change, and Jeff Anderson's *The Lean Change Method* (as mentioned in Chapter 17). These and countless smaller implementations agree on the importance of maintaining an auxiliary pull system in parallel with the main delivery system. These allow small increments of change to be pulled from some kind of backlog (perhaps represented on a *change canvas* or *story map*) and managed visually through their implementation.

There are some interesting differences in emphasis among authors that should be noted. Some like to use the *Kanban Depth Assessment* tool (or similar tool) to prioritize practices that should be implemented or refined; others focus more on dissatisfactions or problems, using these as the driver for change.

The former, practice-driven approach makes good sense when the goal is to achieve a rollout in limited time. For two reasons, though, I lean toward the latter, more problem-driven system: I'm leery of changes that might be seen to be implemented for their own sake rather than to address pre-existing dissatisfactions, and a problem management system is just too useful to be ignored.

Identifying Increments of Change

The nine values represent practices and describe benefits; by implication we can use them to organize the problems they address also (so the apparent dichotomy between practice-driven and problem-driven change needn't be a big deal after all). Let's take one final pass through them and identify key features of effective Kanban implementations.

You can use them to assess where you are, and to prioritize increments of change that make sense in your situation. As you go through this section, give your system a score of 1 through 4 on each numbered feature according to this scale:

1. Our system exhibits this aspect barely, if at all.
2. Our system is somewhat capable of exhibiting this aspect.
3. Our system exhibits this aspect convincingly, for the most part.
4. Our system departs from this only very exceptionally; we manage the consequence when it does so.

Transparency

1. Work items are organized visually by type, state (in some knowledge-discovery activity, waiting in a queue, or some other state), parallel work stream, and class of service.
2. It is clear which items are blocked, and for what reason.
3. To the extent that it matters, it is clear who is working on what.
4. Explicit policies capture shared expectations on work item selection, quality criteria, and so on.
5. The progress of the work and the overall effectiveness of the system are subject to review at a range of cadences, from at least daily (at the standup meeting, for example) up to quarterly and longer.
6. Attention is paid to how progress, demand, and capability are reported externally, both to customers and to the wider organization.
7. Metrics have a clear relationship to the system's purpose.

Even before process changes have been implemented, the introduction of visual management tends to deliver an immediate benefit by making visible the need to make decisions (big decisions, sometimes). The other aspects of **transparency** identified here may take more time to bed down; but they are no less important with regard to the organizational impact they can bring.

Balance

1. Work-in-progress (WIP) is limited such that no individual, activity or work stream is overburdened or is consuming a bigger share of the available effort or shared resources than is appropriate.
2. Work is pulled into and across the system only when capacity is available.
3. WIP limits apply to all work started but not completed; this includes work waiting between activities and between services.
4. The system comfortably accommodates a variety of schedule risk profiles (distinguishing, for example, between date-driven and urgency-driven work) and classes of service.
5. In allocating capacity between competing sources of demand, consideration is given to the needs of all stakeholders and to the overall capability of the system over a broad range of timespans.

Inexperienced practitioners of Kanban seem to worry a lot about the change management aspects of **balance**. I offer two pieces of advice:

◆ Remember that to reduce WIP, more work must be finished than is started. If it's not obvious to people what should be completed first, focus on that issue first. Organize the work visually and with policies, and ensure that it will stay organized through effective feedback loops.

◆ Remember that the level of WIP is both a lever and a symptom. Visualize it; bring its root causes to the surface; expect it to reduce as the process improves.

Collaboration

1. Improvements are framed and structured as experiments and managed visually.
2. Other bodies of knowledge are used as models for improvement (ways of looking at systems, ways to structure change, technical and management practices, and so on).
3. Collaboration is embraced as a source of performance, a driver of improvement, and an antidote to system-generated frustration.
4. The system is open to change from inside (it is self-organizing) as it pursues fitness for purpose.

Remember:

◆ **Collaboration** isn't just "being nice"; neither is it limited to problem solving.

◆ A good response to disappointment is to consider the role collaboration might have played in preempting it.

Customer Focus

1. The delivery workflow is understood to be a process of knowledge discovery, in which needs, possibilities, and capabilities are explored.
2. Upstream of a defined commitment point, work items are managed as options.
3. Downstream of delivery, work items continue to be managed until their utility in the hands of the customer has been validated.

From experience, that last point appears to be crucial. Paradoxically, validation takes place at the end of the delivery process, and yet it's the aspect most likely to bring about collaborative **customer focus** right through the process. At the Lean Startup extreme, it is the process's engine.

Flow

1. Work items are sized and selected to achieve a strong and reliable flow of value.
2. Batches are sized and releases scheduled to maximize overall economic outcomes (not just to minimize delivery costs).
3. Work items of exceptional value or risk are managed appropriately.
4. The system reliably delivers non-exceptional work items with appropriate predictability.
5. Measured end-to-end, time spent in active knowledge discovery dominates time lost to delays (queuing, multi-tasking, blocking) and other kinds of work.
6. Dependencies between work items and on other services are identified and visualized in good time.
7. Work items can be scheduled for release independently of their commitment into the system.

We set high standards for **flow.** Implementations are unlikely to demonstrate many of these aspects very convincingly until they have first focused their improvement efforts on sources of unpredictability and delay and given some serious thought to the true economics of their work.

Leadership and the Leadership Disciplines

1. Leadership is open to all; acts of leadership that bring about change are especially worthy of celebration.
2. There is a shared and ongoing commitment to change, based on an evolving understanding of what we do now and its alignment to purpose from the perspective of all its stakeholders.
3. Evidence of the need for change is kept close to the surface.
4. Change is safe; its downside risks are identified and mitigated (including the risks of changing too slowly or not at all).
5. The potential benefits of change (upside risks) are watched for and nurtured.

6. Change is implemented through agreement; the practices of change and the capability to change are themselves focuses for improvement.

7. Respect is always a given; at times of change, people's attachment to their current roles, organization, and practices is never underestimated.

In your self-assessment, do not exclude the management of changes that originate from outside the system—from senior management or HR, for example. Dealing with these with **understanding**, **agreement**, and **respect** has the potential to demonstrate **leadership** to other parts of the organization.

Visualizing change

Visualizing the Assessment

In Figure 23.1, the geometric mean[65] of the scores in each category are visualized by plotting them in a radar chart.[66] This gives an alternative, values-based realization of the Kanban Depth Assessment tool, one that doesn't stop at Kanban's practices.

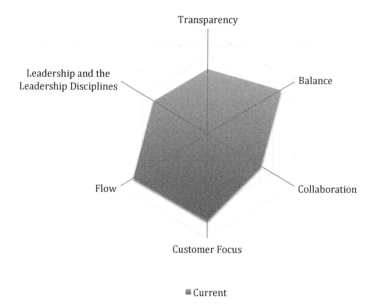

■ Current

Figure 23.1   Kanban Depth, by value

---

65. Relative to the simple arithmetic mean, the geometric mean amplifies the effect of the weaker scores.

66. "Radar chart" is what Excel calls them; you may know them as spider charts or Kiviat diagrams.

It can be useful to visualize historical progress or the desired trajectory also. My friend Ruben Olsen has conducted multiple assessments for the same teams and charted the results over time. Alternatively, you can answer the assessment questions aspirationally, describing where you hope to be in (say) three months' time. Figure 23.2 shows how this approach might be visualized.

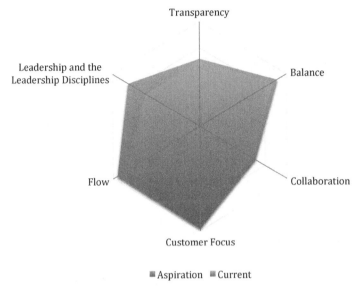

Figure 23.2   Kanban Depth, with trajectory

Managing Change Visually

The challenge is to see some concrete actions brought to fruition in such a way that some momentum is generated. This can feel very much like a delivery process, and indeed it is amendable to being managed with similar tools.

It is unusual to find a significant Kanban implementation these days that doesn't maintain some kind of visual management system in parallel with its main delivery system. These auxiliary systems are devoted to change, problems, out-of-the-ordinary dependencies, and so on.

I co-developed the design of the Problem Board in Figure 23.3 with Kevin Murray of Valtech in 2012. Together and separately we have used several variants of it since.

Problem	Being Sorted		Sorted ☺	
	Daily			Closed
	Weekly			

Figure 23.3 The "Problem Board"

We operate this board as follows:

♦ Anyone is free to add new problems to the input column on the left.

♦ After daily triage and ownership assignment, in-progress problems move vertically between the daily and weekly areas under "Being Sorted" according to the amount of time we wish to devote to discussing them.

♦ Some time after they have been deemed to be "Sorted" and we are sure that they will not resurface, decisions have been logged, and so on, problems move to "Closed."

A more change-focused design is the one shown in Figure 23.4, from Jeff Anderson; it complements his *Lean Change Canvas* (Chapter 17), but could easily be used independently.

I like Jeff's design very much. Its first two columns emphasize **agreement**; the last two emphasize validation (and by extension, **customer focus**). Separating qualitative validation from quantitative verification seems quite smart, too; typically, teams will be happy to confirm behavior changes long before it is possible to confirm any significant performance improvement.

Agree on Urgency	Negotiate the Change	Validate Adoption	Verify Performance

Figure 23.4 A kanban system for "Validated Change"

Not All Change Is Alike

Way back in 2010 or 2011 (which seems aeons ago in Kanban terms) my friend and long-time collaborator Patrick Steyaert provoked me into "*not all work is alike*," a phrase I have used many times since, including in this book.

With "*not all change is alike*," Patrick reminds us of another useful parallel between managing customer deliveries on the one hand and system changes on the other. Referencing Thomas Kuhn's book *The Structure of Scientific Revolutions*,[67] Patrick identifies two axes by which change can be analyzed:

1. Does the change involve reconstructing organizational commitments as it challenges prevailing thinking (revolutionary, *paradigm breaking* change), or does it build on what is there already (cumulative, *paradigm consistent* change)?
2. Is the change imposed from outside (*reactive*) or initiated internally (*proactive*)?

Those two axes suggest a quadrant, shown in Figure 23.5, which Patrick has labelled helpfully, too.

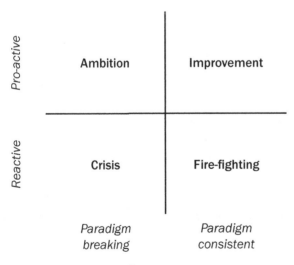

Figure 23.5   Not all change is alike.

---

67. See Patrick's post http://lean-adaptive.com/2014/06/03/not-all-change-is-alike/; his reference is to Thomas Kuhne's *The Structure of Scientific Revolutions* (University of Chicago Press, 50th anniversary ed., 2012)

A few observations:

◆ Kanban is not about fire-fighting or managing existential crises, but if you find yourself needing to react in such a situation, it is possible that Kanban will help.

◆ Introducing Kanban is usually a proactive change, and it is designed to generate further proactive change.

◆ Whether Kanban as a whole, its individual principles, practices, and techniques, or the changes it provokes are paradigm consistent or paradigm breaking depends greatly on where you are now. For example, your organization may already be operating pull systems, or it may find the idea alien.

◆ In the terms of Patrick's quadrant, Kanban is not only about improvement; it is (very much) about ambition.

Neither does Kanban exclude radical change. What it asks is that you understand where you are now before you move forward on the basis of agreement and respect. If you can't or won't do that, then you are—at least as far as Kanban is concerned—on your own.

Closing Thoughts

It seems fitting that I come to the end of the final chapter celebrating other people's work. This is very much a collaborative effort, the work of a community that is open to ideas, embraces change, and cultivates leadership in others.

I'm very glad also that I could finish the book as it started, with values. There were quite a few skeptics when I first laid out the nine; not unreasonably, they worried that it wasn't the method's place (or mine) to dictate the values of their organizations. Some warned me that values are fragile things—touch them, and they disappear!

Used carefully as proxies for practices and benefits, and as tools for organizing thoughts, stories, priorities, and so on, values have proved to be non-threatening, usefully thought provoking, and actually rather robust. To all who recognized these possibilities early on, and to you for sharing in them now, I thank you.

Further Reading

Achouiantz, Christophe and Johan Nordin. 2014. *The Kanban Kick-start Field Guide: Create the Capability to Evolve.* http://leanagileprojects. blogspot.co.uk/2013/11/the-kanban-kick-start-field-guide-v11.html

Anderson, Jeff. 2013. *The Lean Change Method: Managing Agile Transformation Using Kanban, Kotter, and Lean Startup Thinking.* https://leanpub.com/leanchangemethod

Yeret, Yuval. 2013–14. *Pull-based Change Management.* http://yuvalyeret.com/2014/05/27/pull-based-change-management/

❖ Glossary ❖

Activity   In the context of a *workflow*, identified activities are performed on *work items* that are in appropriate *states*; activities often take work items from one state to another. Activities and their corresponding states typically map to the *columns* on a *kanban board*.

Adaptability   The ability to respond to changes in the environment. Note that very stable environments do not necessarily favor adaptability, which makes their inhabitants especially vulnerable when conditions do eventually change. See also *fitness*.

Agenda   For the purposes of this book, an *agenda* (or *agenda for change*) is a compelling call to action based on organizational need. Chapter 10 describes pre-defined "three agendas"; the tactic of constructing an agenda around a small number of prioritized *values* is described in Chapter 23.

Avatar   A movable token that represents a person. On a *kanban board*, avatars can be placed on *cards* to indicate who is working on them; limiting the number of avatars available is a technique to reduce multi-tasking.

Backlog   A list of work items that have not been started. Use this term with care if it is likely to imply stronger *commitment* than is warranted.

Batch, batch size, batch transfer   A group of *work items* that progress through the system (or part thereof) together. Projects often imply large batch sizes at the input and output of the system and large batch transfers between various defined stages in between. A strategy of releasing work into and out of the system in smaller batches has the tendency to reduce *work-in-progress* and *lead times*. See Chapter 15 for an economic treatment. See also *transaction costs*.

Blocked, blocker, blocking issue   A *work item* is said to be blocked when there is some abnormal condition preventing it from progressing. The proximate cause may be referred to as the blocker (often an issue or dependency) and visualized on the *kanban board* against the work item in question. See also *stalled*.

Bottleneck   An *activity* whose *delivery rate* constrains the delivery rate achievable from the system as a whole.

Buffer   A *queue* deliberately placed ahead of a *bottleneck* or other critical activity for the purpose of ensuring a steady supply of work to it.

Capability review   A review of progress and performance held regularly at the departmental level.

Card (or ticket)   A visual representation of a *work item*.

CD3   An acronym coined at Maersk for *cost of delay* divided by duration, where duration is the remaining lead time. The *queuing discipline* known as *weighted shortest job first* (*WSJF*) seeks to minimize CD3. See Chapter 15.

Change agent   Usually an individual, a change agent causes change to happen.

Classes of Service (CoS)   Customer expectations defined broadly for different subsets of the overall workload. They influence selection decisions made inside the delivery system. Different classes of service are typically associated with qualitatively different risk profiles, especially with regard to schedule risk and *cost of delay*.

　　Four generic classes of service are widely recognized: "standard," "fixed date," "expedite," and "intangible." I often describe the first two of these as "urgency-driven" and "date-driven," respectively.

Column, column limit   On a kanban board, vertical columns typically organize work items by *state*, such that they will move rightward across the board as they progress towards completion.

　　Column limits are *work-in-progress (WIP) limits*, constraints on the number of items allowed in a column at any given time.

Commitment, commitment point   Work items are said to be committed when there is a strong expectation shared with the customer that work on them should now proceed.

　　If there is one, the point in the process at which the transition between uncommitted and committed states typically occurs (perhaps the result of a replenishment meeting) is referred to as the commitment point. On a kanban board, this is represented by a line between columns.

　　There may be a second commitment point later in the process, the point at which the decision to release, deliver, or deploy work items is made, singly or in batches.

Cost of delay (CoD)   A measure of the impact of delaying a work item, perhaps to allow another with a higher cost of delay to take precedence. At its most sophisticated, cost of delay is a function of time, and it measures the impact on total lifecycle profits of the affected product or portfolio.

Creative knowledge work   *Knowledge work* that is focused on meeting customer needs through *knowledge discovery processes* and creative problem solving. Many forms of product development and service delivery both inside and outside the technology sector meet this definition.

Cumulative flow diagram (CFD) A stacked line chart that shows the *delivery rate* and *work-in-progress* for multiple *states* or *activities*. See Chapter 17.

Customer lead time A specific *lead time*—the time work items take to go through the system—as experienced by the customer. Typically, this is measured from request to delivery.

Cycle time Use with care! Most often this refers to the *lead time* through the "operational" part of the process (measured from when work starts until it is ready to be delivered), but it may refer to *customer lead time*, or (very differently) to the reciprocal of *delivery rate*.

Date-driven See *classes of service*.

Delivery rate Otherwise known as *throughput* (which is sometimes also a financial measure), it is the number of *work items* delivered from the system (or part thereof) per unit of time.

Dependency *Work items* may be dependent on others (in which case they need careful sequencing to prevent them from becoming blocked), or require attention from other *services* (in which case they may need careful scheduling for reasons of availability).

Expedite Describes the management of work items that must progress quickly through the system at the expense of others. See *classes of service*.

Feedback The upstream flow of information about a product or process.

Feedback loop For the most part, this refers to the deliberate incorporation of feedback in the design of a process so that the resulting product, service, or the delivery process itself can be controlled and improved. Note, however, that not all feedback loops are there by design; some may be hard to identify, and they are not always benign. See Chapter 11.

Fitness In its evolutionary sense, fitness describes how well something is adapted to its environment. In a competitive environment, relative fitness confers advantage. See also *adaptability*.

Fitness criteria The measures that combine to indicate *fitness*.

Fixed Date Describes *work items* that must be delivered on or shortly before a particular date. Typically, the *cost of delay* of such items is highly sensitive to small changes in delivery date around that time, and some active management of the schedule risk is called for. See *classes of service*.

Flow efficiency An important metric, the ratio of *touch time* to *(customer) lead time*. Flow efficiency increases when *work items* are subject to less delay. See Chapter 19.

Histogram A visualization of the distribution of data. In the Kanban context, histograms are particularly applicable to *lead times*.

Input queue   A *queue* placed at the start of the process, holding *work items* that have been selected for processing soon but that are not yet started. Often visualized as a "Ready" *column*. Assuming that a degree of *commitment* is involved, work items in the input queue count toward the overall *work-in-progress (WIP)* in the system, and the input queue is a good place to apply a *WIP limit*.

Intangible   Describes work items whose short-term economic value is hard to quantify but whose presence in the system is vital to its health and performance in the longer term. Often applied to preventive maintenance, experiments, system improvements, and so on. See *classes of service*.

Kanban   An overloaded term, often carrying multiple meanings at the same time. Written in *kanji* (Chinese characters), it means "sign" or "large visual board." Written in *hiragana* (Japanese characters), it means "signal cards" (singular or plural). In technical presentations of the mechanics of *kanban systems*, we usually intend the latter meaning. Informally, it refers to the use of *kanban systems* (visual or otherwise) and the *Kanban Method*.

Kanban board   A visual organization of *cards* (the *kanban*) in a *kanban system*. Typically, boards are arranged in vertical *columns* and (optionally) horizontal *swim lanes*; additional dimensions may be represented by color or other card attributes. Cards move rightward from column to column as the *work items* they represent progress through the system. *Work-in-progress limits* and other *policies* may be represented visually also.

Kanban Lens   A way to understand the organization, operation, and improvement of *creative knowledge work*. Its service-based model offers an alternative view to the more conventional models of project delivery and functional organization. It also distills *STATIK*, the *Kanban Method's* implementation approach. See Chapter 10.

Kanban Method   An evolutionary approach to change described by David J. Anderson in six Core Practices and four Foundational Principles. See Chapter 16.

Kanban system   A *pull system* implemented by limiting the number of *kanban* (*cards*) in circulation.

Kanban system lead time   The *lead time* through the part of the system that is subject to *work-in-progress (WIP) limits*.

Knowledge discovery process   A way to understand the process of *creative knowledge work*. It invites us to acknowledge how little we know at the start of the process, and helps us to identify the various kinds of knowledge that are developed at each stage of the process.

Knowledge work   Work that is mainly about using and developing knowledge (after Peter Drucker). See also *creative knowledge work*.

Lead time   Unqualified, this refers to the time a *work item* takes to progress through a process, and is often used synonymously with the more specific *customer lead time*. See also *cycle time*.

Limit   See *work-in-progress (WIP) limit.*

Little's law   A key result from queuing theory that relates a process's arrival rate (for which we use *delivery rate* as its proxy), *lead time*, and *work-in-progress*. It is named in honor of John Little, who published its first proof in 1961. See Chapter 17.

Model   As per the introduction to Part III, a model may mean: 1) the example of others that can be replicated; 2) some set pattern, template, or routine that gives structure to one's actions or thinking; 3) a way to understanding the world based on a defined set of assumptions; 4) a set of outcomes to be expected as a consequence of those assumptions. The *Kanban Method* contains a specific encouragement to its practitioners to use models to inspire, structure, and guide evolutionary change through collaboration and experimentation.

Multi-level board   A *kanban board* that manages *work items* at multiple levels of granularity (epic and story, for example). See Chapter 22.

Multi-tasking   A condition in which an individual or team has more than one *work item* actively in progress. Controlling the overall amount of *work-in-progress (WIP)* in the system controls *multi-tasking*; the converse is not necessarily true.

Operations review   A divisional meeting, typically held monthly, in which multiple teams share performance data, incident reports, and improvement updates with each other and (ideally) representatives of the customer and the wider organization. An essential *feedback loop* for larger implementations. See also *service delivery review.*

Opportunity cost   The economic benefit forgone when making a choice between alternatives.

Over-burdening   In ordinary usage, it is to give people or processes more work than they can effectively or humanely deal with. More narrowly, it is to maintain more *work-in-progress (WIP)* in the system (or part thereof) than can be sustained.

PDCA Cycle   The canonical improvement cycle known variously as the Deming Cycle, the Shewhart Cycle, the PDCA Cycle, or just plain PDCA. The acronym is short for "Plan, Do, Check, Act"; sometimes you will see PDSA, for "Plan, Do, Study, Act." It describes an experimental (in the scientific sense) way to structure an improvement. See Chapters 3, 11, and 14.

Personal Kanban   The application of *Kanban* to the workload of an individual or small team. In their book of the same name, Jim Benson and Tonianne DeMaria Barry identify two practices most relevant to "*choosing the right work at the right time*": 1) Visualize your work. 2) Limit your work in progress. See Chapter 16.

Policy   An explicit description of expected behavior or process constraint. Policies commonly associated with *kanban systems* include *column*-level "definitions of ready" and "definitions of done." *Work-in-progress (WIP) limits* are also classified as policies.

Portfolio Kanban   The application of the *Kanban Method* to the management of project portfolios.

Proto Kanban   An early-stage Kanban implementation that addresses *multi-tasking* but does not control the amount of *work-in-progress (WIP)* between activities.

Pull system   A broad category of work-management system in which work is started—"pulled" from upstream—only as capacity becomes available. *Kanban systems* are pull systems; the availability of capacity and the ability to pull work is indicated by the gap between current *work-in-progress* and the corresponding *limit*. See also *push*.

Push   The act of placing work into a system or activity without due regard to its available capacity.

Queue   A place in the *workflow* (typically represented by a *column* on a *kanban board*) in which *work items* are held ahead of some later *activity*.

Queuing discipline   The set of *policies* that govern the selection of *work items*. First in, first out (FIFO) and *weighted shortest job first (WSJF)* are two important examples.

Release   The act of delivering to the customer, or the product that will be delivered.

Replenishment, replenishment meeting   The act of populating the *input queue*, and the meeting that achieves it.

Run chart   A chart that shows observations plotted against time. Commonly used to visualize *lead times*.

Safe-to-fail experiment   An experiment conducted that is designed to have only limited impact on the system in the event of failure.

Scrumban   The application of the *Kanban Method* in the context of an existing implementation of *Scrum*. Colloquially, it is *Kanban*, when the "*what you do now*" is Scrum.

Self-organization   The ability of a system to change itself without the intervention of an outside agent.

Service; service orientation   A system designed to benefit its consumers; a paradigm based on services, their interactions, and the results they generate for customers. See also the *Kanban lens*.

Service delivery review (or system capability review)   A weekly or biweekly review or progress-and-performance meeting held at the departmental level. See also *operations review*.

Stalled   A *work item* is said to be stalled when it remains idle because there is no capacity in the system to attend to it. See also *blocked* and *starvation*.

Standard   The baseline *class of service*, applicable to *work items* that are neither *expedited, fixed date*, nor *intangible*.

Standup meeting   A regular meeting (often daily) that is short enough (typically up to 15 minutes) that its participants can remain standing. See Chapter 1.

Starvation   A condition in which people or activities lack work due to inadequate flow from upstream. See also *stalled*.

State   The overall condition of a *work item* that determines where it should be in the system and what *activity* or activities could legitimately be applied to it.

STATIK   An acronym for the Systems Thinking Approach to Implementing Kanban, a recommended approach to implementing and progressing with the *Kanban Method*. Described in Part III.

Swim lane   On a *kanban board*, this is a horizontal lane along which *cards* flow. Swim lanes organize cards into categories; cards typically do not move between swim lanes (implying that they represent categories that are relatively long lived, such as epic- or project-sized *work items*, customers, or *classes of service*).

System capability review   See *service delivery review*.

Ticket (or Card)   The visual representation of a *work item* on a *kanban board*. Often quite literally a card (held on the board by magnets, say), or a sticky note.

Throughput   See *delivery rate*.

Touch time   The amount of time a *work item* spends being worked on, as opposed to waiting in a *queue, blocked*, or *stalled* due to *multi-tasking*. See also *flow efficiency*.

Transaction costs   The overhead costs that drive *batch sizes*. Reducing the truly fixed cost of each transaction (or recognizing that a greater proportion of the transaction cost is in fact variable, a function of *batch size*) allows transactions to be made smaller. See Chapter 15.

Unbounded (infinite) queue   A *queue* that has no *limit*.

Urgency-driven   See *classes of service*.

Values   In this book, values refer to properties that are widely agreed to be desirable, provide some sense of direction (because "more is better," generally speaking), and serve to suggest, organize, or represent helpful practices and artifacts. The nine values described in Part I are abstracted from the *Kanban Method's* core practices and foundational principles. Other schools of thought and different organizational cultures will emphasize different values; values can be useful, therefore, for the purposes of comparison and selection.

Weighted shortest job first (WSJF)   A *queuing discipline* that seeks to minimize *cost of delay* by giving precedence to *work items* that have the largest economic impact in proportion to the remaining time needed to implement them. See Chapter 15.

Workflow   The sequencing of *activities* or (broadly equivalently) of *work item states* that results in products or services being delivered. Workflow tends to cut across considerations of functional structure, though not always optimally. See also *knowledge discovery process*.

Work item   A deliverable or a component thereof (a new product feature, for example). Generally speaking, tasks and activities are not work items in this sense. See also *card*, *workflow*.

Work-in-progress (WIP)   At system level, this refers to work that has been started but not delivered out of the system. In the Kanban Method, we are also interested in where the WIP resides, measuring and controlling the number of *work items* that occupy particular *states* or *activities*. We may also control the allocation of WIP across other categorizations, such as customer, work item type, or *class of service*. See Chapters 2 and 21.

Work-in-progress (WIP) limit   A *policy* that constrains the amount of *WIP* allowed in a given part of the system. Typically expressed as a number—the maximum number of *work items*. WIP-limited systems are *pull systems*. Minimum limits may also be used to trigger *replenishment*.

❖ Appendix A ❖

Deming's 14 Points for Management

1. Create constancy of purpose toward improvement of product and service, with the aim to become competitive, to stay in business, and to provide jobs.

2. Adopt the new philosophy. We are in a new economic age. Western management must awaken to the challenge, must learn their responsibilities, and must take on leadership for change.

3. Cease dependence on inspection to achieve quality. Eliminate the need for massive inspection by building quality into the product in the first place.

4. End the practice of awarding business on the basis of a price tag. Instead, minimize total cost. Move toward a single supplier for any one item, in a long-term relationship of loyalty and trust.

5. Improve constantly and forever the system of production and service, to improve quality and productivity, and thus constantly decrease costs.

6. Institute training on the job.

7. Institute leadership (see Point 12 and Chapter 8 of *Out of the Crisis*[68]). The aim of supervision should be to help people and machines and gadgets do a better job. Supervision of management is in need of overhaul, as well as supervision of production workers.

8. Drive out fear, so that everyone may work effectively for the company (see Chapter 3 of *Out of the Crisis*).

---

68. Deming, W. Edwards. 2000. *Out of the Crisis*. Cambridge, MA: MIT Press.

9. Break down barriers between departments. People in research, design, sales, and production must work as a team, in order to foresee problems of production and usage that may be encountered with the product or service.

10. Eliminate slogans, exhortations, and targets for the work force asking for zero defects and new levels of productivity. Such exhortations only create adversarial relationships, as the bulk of the causes of low quality and low productivity belong to the system and thus lie beyond the power of the workforce.

11. a. Eliminate work standards (quotas) on the factory floor. Substitute with leadership.
b. Eliminate management by objective. Eliminate management by numbers and numerical goals. Instead substitute with leadership.

12. a. Remove barriers that rob the hourly worker of his right to pride of workmanship. The responsibility of supervisors must be changed from sheer numbers to quality.
b. Remove barriers that rob people in management and in engineering of their right to pride of workmanship. This means, *inter alia*, abolishment of the annual or merit rating and of *management by objectives.*

13. Institute a vigorous program of education and self-improvement.

14. Put everybody in the company to work to accomplish the transformation. The transformation is everybody's job.

❖ ACKNOWLEDGMENTS ❖

Only through writing a book myself have I truly been able to understand why the acknowledgements sections of many of my favorite books are so large. The name that appears on the front cover is mine, but I would never have got this far without the encouragement, inspiration, and direct support of many people, not all of whom can be listed here.

First—and most definitely foremost—comes my wife Sharon. The word "amazing" does not begin to describe her! Despite some significant family challenges emerging in the year or so that it took to write and publish this book, she kept me writing, accommodating the 4:30am starts and holding the fort while I was physically or mentally absent. With love, thank you.

From David J Anderson and Associates and our publishing arm Blue Hole Press I must thank David himself, Janice Linden-Reed, Wes Harris, Irina Dzhambazova, Agnes Sellgren, and our former colleague Dragos Dumitriu. Thank you for your individual and shared commitment to this project. Thank you also to copy editor Vicki Rowland—I hope I wasn't too much trouble!

Out of a long list of praiseworthy colleagues I have had the pleasure of working with on recent projects I will single out these: Mark Dickinson, my former MD at Encore International; Kevin Murray of Valtech; Leigh Mortimer and Joanne Clarkson of the Department for Work and Pensions (DWP); Allon Lister and Steve Wood of the Government Digital Service (GDS). Thank you for understanding what I was doing and trusting me to run with it.

These good people provided key feedback on multiple drafts: Greg Brougham, John Clapham, Kevin Murray, Klaus Leopold, Kelvyn

Youngman, Steve Tendon, Wolfgang Weidenroth, and Markus Hippeli. Your hard work and expertise are very much appreciated.

Ben Linders of InfoQ was kind enough to publish a series of articles based on a very early (and horribly dense) draft of what would become Chapter 10. I am indebted to the ever-honest Luke Hohmann for his comments at that crucial time, and of course for investing his precious time later in the foreword of this book.

This book is very much the richer for the recent developments contributed to the Kanban community by David Anderson, Jeff Anderson (no relation), and Patrick Steyaert. For their foundational work I thank Don Reinertsen, Dave Snowden, Chris Matts, and Olav Maassen.

The Kanban community had much to absorb in 2013, a good year! Including several names mentioned already, I highly value the feedback of these people on the values material as it emerged through blog posts, conferences, meetups, discussion forums, private email, and the like: Maria Alfredéen, David J. Anderson, Markus Andrezak, Dimitar Bakardzhiev, Corinna Baldauf, Jabe Bloom, Matthias Bohlen, Royd Brayshay, Pawel Brodzinski, Greg Brougham, Martin Burns, Tom Cagley, Andy Carmichael, Jose Casal, Chris Chan, Fred Engel, Thomas Epping, Zsolt Fabok, Alex Fedtke, Rob Ferguson, Eric Green, Ellen Grove, Torbjörn Gyllebring, Kurt Häusler, Hermanni Hyytiälä, Matthias Jouan, Sigi Kaltenecker, Liz Keogh, Klaus Leopold, Janice Linden-Reed, Simon Marcus, Gaetano Mazzanti, Chris McDermott, Marco Melas, John Miller, Rodolfo Moeller, Carolyn Nelson, Pierre Nies, Frode Odegard, Stephen Parry, Matthew Philip, Ajay Reddy, Arne Roock, Bernd Schiffer, Karl Scotland, David Shrimpton, Mattias Skarin, Dave Snowden, Patrick Steyaert, Jim Sutton, Jon Terry, Simon Thomas, Björn Tikkanen, Dave White, Eric Willeke, and Yuval Yeret.

❖ ABOUT THE AUTHOR ❖

Mike Burrows is UK Director and Principal Consultant at David J Anderson and Associates (djaa.com). In a career spanning the aerospace, banking, energy, and government sectors, Mike has been an IT director, global development manager, and software developer. He sits on the management board of Lean Kanban University (LKU), with whom he is an Accredited Kanban Trainer (AKT) and Kanban Coaching Professional (KCP). He speaks at Kanban-related events around the world, blogs at positiveincline.com and tweets as @asplake and @KanbanInside.

❖ GET IN TOUCH ❖

Don't hesitate to contact the author directly at mike@djaa.com if you have a correction or comment on the content of this book. Mike Burrows is available for speaking, consulting, training, coaching, and interim management. For service-related enquiries, please contact sales@djaa.com.

Twitter

http://twitter.com/asplake
http://twitter.com/KanbanInside

Blog

http://positiveincline.com/
http://www.djaa.com/blog

With Our Partners

Kanban Training Classes
http://edu.leankanban.com

Lean Kanban Conference Series
http://conf.leankanban.com/

Twitter

http://twitter.com/LeanKanban
http://twitter.com/LeanKanbanU

For further information about Lean Kanban University certified kanban training or consulting in your area, please contact info@leankanban.com.

Teaching / Coaching Toools

Kanban Knowsy
http://positiveincline.com/index.php/kanban-knowsy/

Featureban Game
http://positiveincline.com/index.php/featureban/

Kanban Values Exercise
http://positiveincline.com/index.php/values-exercise/

Values-based Assessment Tool
http://positiveincline.com/index.php/values-based-assessment-tool/

GetKanban board game
http://getkanban.com/

Kanban Community Connections

Kanban Discussion Group
https://groups.yahoo.com/neo/groups/kanbandev/

Lean Kanban LinkedIn Group
https://www.linkedin.com/groups/Lean-Kanban-4794591

Limited WIP Society - local community groups
http://limitedwipsociety.ning.com/

Meldstrong Community Stories
https://www.meldstrong.com/sites/kanban

❖ INDEX ❖

Kanban and, 130–131
kanban and, 122–123
Lean Product Development
(LPD), 126–128
models and, 84
TPS and, 124–125
lean change canvas, 159, 166–167
lean decision filter, 73n29
Lean Kanban University (LKU), 150
lean product development (LPD),
126–128
lean startup model, 128–129
Lean UX, 129
Lean/Agile hybrids, 129–130
learning
double-loop (Argyris), 93
set-based learning, 127
single-loop, 93–94
learning organization, 85, 91, 94
leverage points, 85–87
life cycle profits, 140–141
Little's Law, 159–162, 235
LPD. *See* lean product development

M

MECE (mutually exclusive, collectively exhaustive), 165–166
meetings, 3, 9–11
mental models, 94
mentored change agent, 66–67
metrics-based feedback loops, 11–12
models, 83–84, 128–129, 235
multi-level board, 235
multi-level design, 208–210
multi-tasking, 235

N

needs of customer, anticipating, 36
noble patterns, 75–76

non-instant availability (NIA)
resource, POOGI, 100
nonlinear systems, 88
nonlinearity, 85
numbers, leverage points, 87

O

obvious domain (Cynefin framework), 89
open-source development, 113
operations reviews, 11, 235
opportunity cost, 137, 140, 235
organization
bottom-up (workflow), 194–195
self-organization, 6, 13, 17, 63,
69, 71, 95, 110, 201, 212,
222, 236
over-burdening, 17, 235

P

pair negotiation (XP), 112
pair programming (XP), 42, 112
paradigms, leverage points, 87
parallel states, design, 204–205
Pareto charts, 186
Pareto Diet, 134–139
patterns. *See* anti-patterns; noble
patterns
expand-collapse, 208–210
PDCA (plan, do, check, act), 28–29,
235
safe-to-fail experiments, 90
peer review, collaboration and,
26–27
performance (J curve), 57–58
Personal Kanban, 152–153, 235
priority sieve, 34
personal mastery, 94
poka-yoke, 122

Made in the USA
Columbia, SC
10 October 2020